SPACE
ENTERPRISE

SPACE ENTERPRISE
Beyond NASA

DAVID P. GUMP

PRAEGER

New York
Westport, Connecticut
London

Library of Congress Cataloging-in-Publication Data

Gump, David.
    Space enterprise : beyond NASA / David P. Gump.
       p.   cm.
    Bibliography: p.
    Includes index.
    ISBN 0-275-93314-8 (alk. paper)
    1. Space stations. 2. Space industrialization. 3. United States.
National Aeronautics and Space Administration. I. Title.
    TL797.G86   1990
    333.9'4—dc20   89-31463

Copyright © 1990 by David P. Gump

Library of Congress Catalog Card Number: 89-31463
ISBN: 0-275-93314-8

First published in 1990

Praeger Publishers, One Madison Avenue, New York, NY 10010
A division of Greenwood Press, Inc.

Printed in the United States of America

The paper used in this book complies with the
Permanent Paper Standard issued by the National
Information Standards Organization (Z39.48-1984).

10 9 8 7 6 5 4 3 2 1

To my wife Wanda, whose love and support for me and our son Chris has endured despite heavy burdens. She's always worked to help us, her friends, and her community.

Her spirit is proof that by reaching out to help each other, we humans are capable of greatness. I hope that space exploration will be a way to demonstrate this fact. On a new frontier there need not be a loser to balance every winner—we can cooperate to improve everyone's life.

Contents

Introduction: The Overdue Frontier   1

1   The Shuttle: Pushed Beyond Its Limits   9

2   The Birth of a New Industry   27

3   Rockets and Spaceplanes of the 1990s   47

4   The $5,000 Ticket to Space   77

5   Space Communications: Ready for New Growth   91

6   Wonder Drugs and Electronics from Orbital Factories   109

7   3M: Looking for Profits in Zero-Gravity   125

8   Orbital R&D for Earth-Side Problems   135

9   Spotting Earthly Treasures with Orbital Cameras   149

10   Prime Real Estate: The New Space Stations   163

11   The New Trade Route: The Moon, Mars, and Low Earth Orbit   183

Contacts for Further Information   205

Index   217

SPACE
ENTERPRISE

Introduction:
The Overdue Frontier

Americans hoped for a new frontier when John Glenn sailed into orbit in 1962. But more than 27 years later, the frontier is still remote. No American colonies are spreading across the sky in orbiting space stations, or occupying Moon bases. The only space workers are a handful of federal astronauts. For the rest of us, the space age was a flop. We weren't invited to the party, and it quickly ran out of steam, anyway.

Now a new space age is beginning. It's driven by private initiative — citizens and corporations creating their own space projects. National governments still dominate space, but for the first time, private programs are under way.

The *Challenger* shuttle mission erupted into tragedy just as this trend began. Many firms, with plans tied to the shuttle, were hurt as they lost access to space. But the tragedy also freed space enterprise from the need to move in lockstep with the National Aeronautics and Space Administration (NASA). Before the disaster, companies almost *had* to win the specific blessing of NASA. Investors and customers rejected "rogue" companies that couldn't secure NASA approval through contracts or cooperative agreements. The tragedy shattered this unwritten rule, a pivotal step because NASA officials are responsible for spending tax dollars and never could allow the gambles and deal making needed for successful free enterprise.

NASA's current troubles began decades ago when President John F. Kennedy ordered it to win the Moon race. Costs weren't important — a $20 billion splurge ensured that Americans were the first lunar explorers. But after the first few visits, the tremendous expense of Moon visits proved too costly a venture for Congress to continue.

Seeing Project Apollo's fantastic price tag, Americans assumed that only the government could afford space programs. The extravagant cost of space exploration seems self-evident: Millions of pounds of fuel go up in flames each time the shuttle flies; clearly, only a government can afford that kind of fireworks display.

This assumption is wrong. The cost to reach space isn't fixed at enormous levels by the fuel required. Propellants amount to only *1 percent* of each shuttle flight's cost. The bloated expense of NASA's space shuttle stems from huge payrolls and overly complicated equipment. Huge payrolls and complex hardware are no accident—they are the inevitable product of a government bureaucracy that must meet political goals rather than economically efficient goals. No other outcome was possible.

THE NEW SPACE AGE

NASA technology and methods in the Apollo era led the world in innovation. The agency had the cash to find a solution to any problem. The space shuttle didn't have such luxury. NASA managers had to make do with whatever money they could coax out of Congress. The more promises they made for the shuttle, the more money they could get. So the shuttle was promised as the answer to every possible space job: commercial cargo transporter, military satellite launcher, orbiting repair station, astronomy platform and science lab carrier. The shuttle was designed to be large enough to carry modules for a space station or an eventual Mars expedition. Every potential use was accommodated in hopes of attracting more supporters in Congress.

The result of too many jobs and too little cash was a space shuttle with dangerous faults and horrendous operating costs. Chapter 1 shows how flawed the final design actually was. The cash shortage also caused a decline in NASA's technical leadership. Development work had to be stretched out and slowed down so much that NASA designs began to lag behind the commercial world. Tight budgets also shut NASA's doors to most new college graduates after Apollo, an outcome that eventually isolated the agency from the mainstream of university research. With the average employee now 46 years old and facilities run down from a decade of tight budgets, NASA often can't keep up with leading edge technology.

The shuttle program constantly scrounged for cash, forcing other parts of NASA to survive on crumbs. NASA research into manufacturing in space or future low-cost vehicles just didn't have the money to be very active. Companies and citizens in the outside world, by contrast, have been fighting in a competitive world to keep ahead of the rush of new technology. The outsiders now have tools and inventions that make them serious players in the space effort even though they lack NASA's billions.

The American Rocket Company (AMROC) is an example. AMROC has test-fired the engines for its new commercial cargo rocket via a Macintosh computer-controlled process. The Macintosh screen displays a diagram of all the valves and cables leading to the engine. Each procedure — opening a spigot to start fuel flowing, for example — instantly appears on the screen. When everything is ready, a technician simply clicks the Macintosh mouse, and the engine thunders into action! A new entrepreneurial space age is possible partly because personal computers can outperform the mainframes NASA used in Apollo. (IBM says its $10,000 top-of-the-line PS/2 personal computer is equal to the power of the $3.4 million System 370 mainframe of the early 1970s.) This puts design and computational power for space projects into an affordable range.

The other prime factor in the new space age is competition. The private sector, for all its faults, still gets far more bang for a buck than a government agency. Look at recent innovations in mail services. Federal Express revolutionized overnight mail delivery, not the U.S. Postal Service. And United Parcel Service virtually took over package delivery with the novel idea that parcels should arrive quickly, inexpensively, and in one piece. These upstarts dominate key specialties in mail delivery even though the U.S. Postal Service had all the advantages in experience, size, and national presence.

Space companies already are demonstrating the same advantage. For example, the military is creating the NAVSTAR satellite system to provide precise navigation data to all its units. At the same time, the Geostar Corporation is building a civilian system based on different technology. Geostar's founder estimates that the Pentagon system may cost $30 billion by the time it is finished in the mid-1990s, whereas Geostar will cost only $200 to $300 million — a 100:1 ratio of government dollars to private dollars to achieve much the same result.

To be fair, the Pentagon's NAVSTAR system does many things in addition to navigation. It watches for nuclear explosions and sends out timing signals that will be essential to the government's secret coded communications. These added tasks are typical of almost all government projects — nothing ever stays simple. Private operations can "just say no" to proposals to make a project more complicated. Most executives know that any new venture is exceedingly hard to accomplish and must be kept simple to have any hope for success.

This specialization — concentrating on a single goal — is almost as important as competition in cutting space costs.

High launch costs have always been the chief obstacle to space development. The new private rocket firms won't eliminate launch costs as a problem, but they will trim them back to a manageable size. The space shuttle's costs are enormous—about $6,000 per pound to put people and cargo into low Earth orbit. That's actually higher than Apollo-era launch costs, even after adjusting for inflation. The private rocket companies expect to bring costs down to $1,000 or $2,000 per pound by 1990, and possibly down to $500 per pound by 1995. That's not cheap enough to start a tourist boom, but it's low enough that companies can seriously consider building small factories and research labs in orbit.

BUT IS SPACE GOOD FOR ANYTHING?

The Apollo program resembled a diet soda: a great first impression followed by a disagreeable aftertaste. Winning the Moon race was nice for the national ego, but once we got there, it seemed a little pointless. The Moon turned out to be worse than a desert. Gray rocks and gray dust had all the romance of an abandoned cement plant. There was no hint of anything useful to humankind.

Taxpayers certainly saw nothing worthwhile about the Moon, especially after color photos in *Life* magazine were just as bleak as the black and white TV coverage. But now, 17 years after the United States abandoned the Moon, we've discovered ways to make all that gray dust productive. The Moon actually is a storehouse of materials that will be very useful in building a space civilization.

Locked up in all that dust are huge amounts of oxygen, raw glass, iron, aluminum, and silicon. The oxygen is precious in space for fuel as well as for breathing. The glass can be woven into new composite building materials similar to fiberglass. The iron and aluminum can be made into machinery and tools or turned into a powder for rocket fuel; and all it takes to harvest the iron is dragging a magnet across the dust.

These supplies eventually will be exported to low Earth orbit where space stations will need them. Getting supplies from the Moon instead of Earth makes sense because the Moon's gravity is so weak. Boosting materials from the Moon takes only 5 percent of the energy needed to lift supplies from the Earth. The low gravity of the Moon will make it the 7-Eleven of space—a convenient stop for simple supplies when one doesn't need the high technology of Earth.

Asteroids that wander across the Earth's orbit could serve the same function. Some may even be rich in strategic metals such as chromium and platinum that we now must buy from the Soviet Union and South Africa. Nobody's explored these close-passing asteroids yet, so their true potential is still speculative. But we do have samples from the Moon, and we know it can be useful.

INDUSTRIAL PARKS IN ORBIT

Space stations and colonies in low Earth orbit will have two main roles. First, they will be assembly points for further exploration. Second, they will become research and manufacturing centers. The two roles result from the fact that space is a dual frontier: It's new real estate to explore, *and* it's a laboratory where zero gravity warps the laws of physics. Dozens of new industries will exploit zero-gravity science, producing medicines and electronic devices that can't be made on Earth. The first space factories already are under construction and will be carried into orbit when the shuttle works off its backlog of high-priority military and space science flights. The space factories are tiny—often about the size of an office copier—but they are just the first step in a series of larger units.

Research labs also will be important, rivaling the space factories as orbital employers. Scientists working in zero gravity will be able to solve some research problems faster than Earth-bound workers, for the simple reason that removing gravity also removes complicating variables in many types of experiments. Scientists can go right to the heart of a problem.

Consider the very basic step of combining two chemicals in a test tube. Often, one is heavier than the other and sinks to the bottom. Shaking or stirring the chemicals to keep them mixed complicates the experiment: Was the result caused by combining the chemicals or by shaking them? In zero gravity, the heavier chemical doesn't fall to the bottom of the test tube, and thus no shaking or stirring is required. Whatever reaction takes place is strictly the result of combining the two chemicals.

Cancer research is another example where zero-gravity may simplify research. Today, Earth-bound workers create new cancer drugs and then test them on herds of mice for safety and effectiveness. Successful drugs then go through another round of tests on primates such as chimps and apes. Only then can the drugs be given to human volunteers for a year or two of clinical trials. A decade or longer may elapse between discovery and market delivery of a cancer drug.

Space labs could slash testing to a few months by providing a missing tool: live human cancer tumors growing outside the body. With human tumors available, cancer researchers could skip the mice and monkey tests. Complex human tumors can't be kept alive outside the body in normal gravity, but researchers believe they could be sustained in zero gravity. Promising drugs could be injected right into their real target. The cost of testing drugs would plummet, hundreds more compounds could be screened, and cancer survival rates would rise dramatically.

Orbital labs will provide many more of these research shortcuts. Some will help solve Earth-bound problems, and others will find new ways to use orbital factories. The potential exists for orbital colonies to pay their own way by producing both knowledge and products. The hardest step will be the first—actually building the first permanent American space station.

This won't be easy for the United States because space stations are a Soviet specialty. The Soviet Union started in 1971 and had two space stations orbiting when the *Challenger* destructed. Soviet crews shuttled back and forth between the two stations at the same time the U.S. space program was grounded by the *Challenger* and other disasters.

Americans have lofted only one space station—*Skylab*—in 1973. Various crews logged less than six months work in *Skylab* before NASA ran out of money for new Saturn rockets to get them there. It crashed back to Earth in 1979, raining debris over Australia and the Indian Ocean. The next U.S. space station is still being designed. With luck, NASA may put it into operation by 1996.

Obviously, a 27-year lag between walking on the Moon and putting up our first permanent space station shows that NASA moves in slow motion. Many companies have recognized this truth and the opportunity it presents for nimble-footed private initiatives. Some are creating automated space platforms to precede the official space station. Others have privately developed science specialities that may turn a profit in orbit. And several, like AMROC, are developing the low-cost boosters needed to bring real growth to the space economy.

Nonprofit efforts also are spreading. One of the most successful is the International Space University, which gathered 104 top graduate students from 20 countries for its first eight-week summer session in 1988 at the Massachusetts Institute of Technology. It was followed in 1989 with a summer session at the Universite Louis Pasteur in Strasbourg, France.

Three individuals—Peter Diamandis, Todd Hawley and Bob Richards—created the International Space University by leading a fund-raising drive that netted $1 million. The students heard lectures and participated in the design of an international lunar facility for research and industry.

The International Space University is just a single example from dozens of space camps, lecture series, and symposia created by space activists. They complement the commercial initiatives under way around the world. Space development will soon move *beyond NASA* as these private efforts multiply. Some pioneers will make fortunes and create new business empires. The competition will finally move space into high gear.

The outlines of a second space age are clearly visible today—complete with new careers and new opportunities for investors. The next chapters show what projects are under way now and how we'll prepare for the coming space economy.

The Shuttle: Pushed Beyond Its Limits 1

The night of January 27, 1986, the space shuttle *Challenger* sat on a Florida launchpad in a chill wind from the northwest. The huge external tank, the two solid rocket boosters, and the orbiter grew colder by the hour. Crews labored to ready the *Challenger* for a launch the next morning, just 24 hours after high winds forced cancellation of the previous attempt.

The weather was causing problems for the upcoming launch, some seen, some invisible. The predicted hard freeze was a threat to the water lines on the launch tower next to the orbiter. Shuttle planners hadn't put insulation on the water pipes feeding the fire hoses and eyewash showers. Kennedy Space Center technicians applied the only high-tech solution available: They opened the faucets to keep the water moving overnight.

Out in the Atlantic Ocean, two ships struggled to get into position to retrieve the solid rocket boosters after they burned out and fell back to Earth. But in the Atlantic 120 miles east of the space center, the weather was even harsher than on shore. Waves 25 to 30 feet high battered the recovery ships so heavily they abandoned course. The ships turned to point into the wind and simply hung on in "survival mode."

In northern Utah, two Morton Thiokol engineers worried about the unseen effects of the frigid night. They believed the solid rocket boosters weren't reliable in the historic cold snap predicted for early morning. Temperatures at dawn were forecasted to be only 23 degrees. The engineers, Roger Boisjoly and Arnold Thompson, argued against a launch in a three-way conference call with their superiors and NASA engineers in Florida and Alabama.

The boosters had a weak point, and the engineers worried the cold would push them to the breaking point. Each booster was built in segments, and the joint between segments was a trouble spot. Recovered boosters showed erratic damage to huge rubbery O-rings that supposedly sealed the joints. Jets of rocket exhaust sometimes leaked around or tore through the

O-rings. The worst damage had been done on a shuttle launched the previous January—the 53-degree weather had been the coldest experienced to date.

Thompson and Boisjoly told the NASA engineers and managers that the O-rings were too tender to fly in freezing weather. Because the damage had been so extensive at 53 degrees last January, they saw tremendous risk in launching at a colder temperature. NASA's experts were not happy with that advice.

"I'm appalled at your recommendation," said one of the NASA officials, George Hardy. He was the deputy director for science and engineering at the Marshall Space Flight Center, the NASA center that controlled Morton Thiokol's contract for the boosters.

"My God, Thiokol, when do you want me to launch? Next April?" complained another NASA manager, Larry Mulloy. He was the contract manager for the boosters.

By 10:30 P.M. Eastern Standard Time, the Thiokol group knew that NASA was extremely upset. They called for a time-out and killed the microphones in Utah.

Jerald Mason, a senior Thiokol vice president who had been silent during the engineering debate, announced a change in approach. The question of whether to launch would have to be a management decision, not an engineering decision.

Mason and three other vice presidents were seated at a front table in the room. Thompson, an engineer who supervised the booster structures, got up and went to the head table to argue with the vice presidents. Thompson put a pad down in front of the managers and tried to sketch out, again, the joint's problems. Partway through his explanation, Thompson realized the executives weren't interested and stopped.

Boisjoly, the staff engineer in charge of the O-rings, grabbed his photos of past O-ring damage and took them to the vice presidents' table. Last January's flight showed black soot on the O-ring over a tremendous arc, almost one third of its total length. The vice presidents weren't impressed, because flights in warm weather had some O-ring damage, too, although not as much.

After the engineers gave up, Mason turned to Robert Lund, the vice president for engineering at Thiokol. Mason told Lund to take off his engineering hat and put on his management hat. All the vice presidents knew, from their management positions, that NASA had become a very unhappy customer. NASA wanted absolute proof that it was unsafe to fly, and Thiokol didn't have it—mostly because NASA and Thiokol simply hadn't spent enough money on testing.

The Thiokol vice presidents resumed the conference call and gave NASA permission to launch.

MORE DANGERS ON THE PAD

Around midnight, the external tank began filling with supercold liquid hydrogen and liquid oxygen. The external tank was joined to the boosters with struts, and the bottom strut anchored to the boosters just a few inches below one of the tender O-ring joints. All through the night, heat seeped away from the O-ring joint through the strut to the supercold external tank.

As it cooled, the external tank contracted. The struts began pulling against the boosters, which were bolted to the launch platforms. A force of 190,000 pounds gradually built up as the shrinking external tank strained against the solid boosters.

The right-hand booster suffered from an additional flaw. The segments were shipped from Thiokol's Utah plant on their sides, and the tortuous rail trip across nearly an entire continent had bent them. They weren't round when they arrived in Florida. Near the super-cold strut, the bent segments almost touched, leaving no room for the O-ring to function.

The joint was designed to seal when the first blast of rocket exhaust pushed the O-ring into place. The gap between the segments had to be just big enough to allow the O-ring to move into a snug fit.

So the joint was triple doomed: the freezing northwest winds, the tremendous cold of the liquid hydrogen transmitted from the external tank, and the squashed shape of the segments caused by the trip from Utah.

Launch crews had other worries. The external tank's nose cone was getting too cold. The nose cone needed to be at least 40 degrees, so that sensors could accurately take readings on the pressure inside the tank's two propellant sections. Built-in heaters were running full blast but couldn't keep the nose cone warm.

Tapes of Kennedy Space Center officials reveal how the nose cone problem was handled:

Launch director: What are they [the nose cone readings] today?

Engineering director: It's been down as low as 10 basically.

Launch director: Wow!

Later, the officials discussed what to do with this key part—the nose cone was running 30 degrees colder than standards allowed. Faulty readings on the propellant pressure could shut down the main orbiter engines and force an extremely dangerous emergency landing.

Launch director: Okay. The only outstanding item we have right now is the one waiver on the cone temps.

Engineering director: Okay. It looks like we probably could say about ten degrees and be okay on that one.

Launch director: Okay. We'll use ten degrees then.

The officials didn't log any written explanation why 10 degrees was "okay" that morning when the previous 24 launches had a 40-degree limit.

A few hours before dawn, another problem was obvious: Letting water run to keep the pipes from freezing had backfired. Drains had frozen, and winds had sprayed the water all over the launch tower so that it resembled an ice sculpture. There was danger to the crew from a sheet of ice covering the floor leading to their emergency escape route. And there was danger to the orbiter from the ice chunks that would shake loose on lift-off. The orbiter tiles are extremely fragile — they can be damaged by a mere fingernail scratch — and the launch tower ice could smash so many tiles that the orbiter would burn up on reentry.

Kennedy officials talked with the ice team leader over the radio system about the predawn launch tower inspection:

Engineering director: You feel comfortable with what you see out there, Charlie?

Ice team leader: We have a lot of ice, if that's what you mean. I don't feel comfortable with what's on the FSS [Fixed Service Structure, the launch tower].

Engineering director: Then what choices we got?

Ice team leader: Well, I'd say the only choice you got today is not to go. We're just taking a chance of hitting the vehicle.

Launch director: You see that much ice?

Ice team leader: Well, the problem is that we have a lot of icicles hanging, you know, even on the west side of the FSS here, which is only 60 feet or more from the orbiter wing. And I'm sure that stuff is going to fall off as soon as the acoustics get to it. And if we're worried about that little bit of ice that comes off [the fuel vent arms], what we have over here is considerably more than that, you know — it's a hundredfold.

The NASA managers continued to argue with their ice inspection expert, doubting his evaluation that all the ice he saw could really be bad enough to stop the launch. They asked Rockwell International, builder of the orbiter, to run computer simulations on how the ice chunks would fall.

The request went to Rockwell's plant in Downey, California. Tapes show how Rockwell engineers at Kennedy and Downey viewed the problem:

Downey: Looks bad, eh?

Kennedy: Ice does look bad, yeah. . . . The big concern is gonna be the mass of ice that is on the FSS, from the 235-foot level all the way down to the MLP [mobile launcher platform]. Every platform had water running on it all night, and some of the close-ups of the stairwells looks like, uh, something out of *Dr. Zhivago.* There's sheets of icicles hanging everywhere. . . . The initial walkdown said icicles up to two feet long by an inch in diameter. . . .

Downey: Sounds grim.

Kennedy: The big concern is that nobody knows what the hell is going to happen when that thing lights off and all that ice gets shook loose and comes tumbling down and—what does it do then? Does it ricochet? Does it get into some turbulent condition that throws it against the vehicle? Our general input to date has been that there's vehicle jeopardy that we're not prepared to sign up to. . . .

Downey: Okay. We didn't see this when we had icing conditions before?

Kennedy: No, and they didn't run the showers all damn night before. They ran the showers this time, and ran 'em pretty heavily, by the look of it; the drains froze up and they all overflowed

Downey: Oh. . . .

Kennedy: And I guess nobody watched all night, or if they did, they didn't say anything. But is John [Peller, an engineering vice president] in yet?

Downey: No.

Kennedy: Okay. . . . somebody at his level needs to get in and try to get up to speed as fast as they can. They're going to be looking for a final position from Rockwell here very shortly. . . . We're probably going to be the forcing factor on this thing. . . .

Rockwell tried computer simulations on the ice breakup and found three different ways the ice could damage Challenger. The damage wasn't certain, but it was very possible. NASA engineers, running their own computer projections, said the danger was acceptable. Just before a 9:00 A.M. meeting on ice conditions, Rockwell vice president Robert Glaysher at Kennedy compared notes with the company's chief engineer at Downey:

Engineer: We still are of the position that it's still a bit of Russian roulette; you'll probably make it. Five out of six times you do, playing Russian roulette. . . .

Glaysher: OK. Our position fundamentally hasn't changed. We'll just go in now; we got a 9:00; we'll go in and express it. I'll let you know what happens.

Rockwell executives told NASA's top shuttle manager, Arnold Aldrich, at the 9:00 A.M. meeting that Rockwell couldn't guarantee a safe launch with all the ice on the launch tower. They couldn't guarantee it would fail, either, so Aldrich decided to press on.

The *Challenger's* main engines ignited at 11:38 A.M. Chunks of ice began falling in patterns not predicted by the computer projections. Six seconds later, the solid rocket boosters roared to life. At this point, the temperature of the crucial joint was only 16 degrees.

Explosive bolts erupted, freeing the boosters from the launch platform. The bolts also released the tremendous tension between the shrinking external tank and the boosters — and the sudden release set the joint vibrating.

Hot propellant gas from the solid rocket immediately burned past the O-ring at ignition. Small clouds of smoke near the bad joint, unnoticed at launch, showed up on TV tapes of the launch. For the next three seconds, as the joint vibrated, hot rocket exhaust escaped through the side of the booster.

Then, as the vibrations damped out, burned insulation and rocket fuel built up around the leak and actually sealed the joint. With luck, the boosters could have fired normally for the rest of the launch.

"Looks like we've got a lotta wind here today," was the comment from pilot Mike Smith, 19 seconds after lift-off. In fact, the *Challenger* flight hit more wind turbulence than any previous shuttle mission.

"Okay, we're throttling down," announced Commander Dick Scobee at 43 seconds. Every shuttle flight eases off on engine power at this point, as the orbiter passes through the period of maximum stress and pressure.

At 57 seconds, Scobee announced, "Throttling up."

At 58 seconds, the *Challenger* again hit heavy air turbulence. The shaking broke loose the burned material clogging the joint, and the *Challenger's* luck ran out. A flame began shooting out of the booster.

"Woooohooo!" shouted Smith, feeling the orbiter shake from the pressures.

At 61 seconds, the blowtorch of hot booster gasses burned through the external tank.

"Thirty-five thousand [feet] going through one point five [indicating 1.5 times the speed of sound]," Smith reported at 62 seconds.

The entire external tank ruptured, spilling liquid hydrogen. At 73 seconds, Smith said "Uh oh," and Kennedy lost all contact with the orbiter. *Challenger* broke apart, with pieces sailing higher for 25 seconds, then plunging down to the Atlantic.

The crew compartment stayed in mostly one piece; several astronauts turned on their emergency air supplies. The air packs were created for ground emergencies when astronauts might have to go through smoke to leave the orbiter. They weren't designed to replace cabin air if the orbiter lost pressure in an accident. Whether the *Challenger* lost pressure isn't known; the astronauts may have been conscious as they plunged toward the Atlantic. The crew cabin hit the water at 200 miles an hour, killing everyone.

NASA never allowed the Brevard County medical examiner, Dr. Laudie McHenry, to conduct autopsies. When NASA later presented Brevard County authorities with death certificates to sign, they refused because they'd never seen the bodies. NASA wants the public to assume that the astronauts died instantly.

CHALLENGER AND SHUTTLE ECONOMICS

NASA's decision to launch *Challenger* seems incredible, considering how many factors argued for a delay:

—Huge amounts of ice ready to break free and ruin the orbiter's heat-shield tiles.

—Important fuel sensors on the external tank falling 30 degrees below the previous minimum standards.

—Trouble-plagued booster joints that had seen their worst problems in a cold snap the previous January.

—Booster recovery ships so battered by an Atlantic storm that they'd gone into survival mode. The ships were fleeing the recovery zone, and NASA expected to lose close to a million dollars in booster equipment as a result.

Why was NASA so determined to launch *Challenger* despite these problems? The answer is shuttle economics.

NASA needed 24 flights a year from its fleet to bring down average costs—otherwise, Congress and the public would declare the shuttle program a failure. Lower costs, after all, were the only reason NASA spent $21 billion to develop the shuttle.

On the tragic morning of January 28, the space agency was far away from its 24-flight goal. It had completed only nine flights in 1985. The new year *had* to run on time; the system had been declared "operational" in 1982, and by 1986, NASA was running out of excuses.

The agency set a goal of 15 flights for 1986. Three of these missions absolutely, positively had to launch on time: a March flight to observe Halley's comet and two deep-space probes in May. Delays would ruin the once-every-76-years chance to observe Halley, and another launch window for the Jupiter probe was years away.

Managers preparing for the launch of *Challenger* knew the 1985 record simply wasn't acceptable to NASA or Congress. The nine launches cost an average of $360 million. This figure was far more expensive than competing European launch services, or even the cost of NASA's old expendable Saturn rockets.

The space agency couldn't sell services to anyone at the actual 1985 per-flight cost of $360 million. Instead, NASA set prices based on the "expected" lower costs of the shuttle when it reached the 24-flight level. The projected savings allowed NASA to book customers for only $74 million per flight. The problem was that flying the nine missions in 1985 already had put enormous strains on the shuttle. Increasing the launch rate to 24 flights was a fantasy but no NASA official or contractor wanted to be the first to admit defeat.

The 1985 actual costs worked out to $6,000 per pound of cargo lifted to orbit. The throwaway Saturn rockets used in the Apollo moon program were cheaper. Each Saturn heaved 200,000 pounds into low Earth orbit for $300 million. That's $1,500 per pound in 1972 dollars and roughly $3,800 in current dollars. NASA would be disgraced if the shuttle program stuck at nine flights a year—a vehicle it designed expressly for low costs would be revealed as a failure. Even worse, NASA probably would be forced to stop offering low-ball prices. Fair shuttle prices would leave room for private competitors to challenge the agency's space monopoly.

SPACE CAN BE AFFORDABLE

Space transportation doesn't have to be expensive. The laws of physics do not require billion-dollar price tags for space projects. The high costs of NASA programs are instead dictated by the laws of government: Agencies with monopolies are likely to do things the safe, expensive way. And to get congressional approval, agencies often skimp on up-front investment costs and suffer later with high operating costs. Any such bureaucracy—whether it's NASA or the U.S. Postal Service—will operate with inflated costs.

The flame and thunder of a space launch seem terribly expensive—we naturally think only a government could afford it. But the flame and thunder from burning massive amounts of rocket fuel is a tiny part of launch costs. The $2 million expense of shuttle propellants is less than 1 percent of a mission's overall costs. What *does* run up shuttle costs is an enormous payroll (more than 24,000 NASA and contractor workers to operate and refurbish the fleet) and a vehicle designed by committee.

The shuttle design is a mass of compromises. NASA decided it needed a big, reusable vehicle, and such vehicles make economic sense only when flown dozens of times a year. Finding customers for all those flights meant the shuttle would have to serve every possible need—deep-space science missions, communications satellite launches, microgravity research, and eventual service as a space station ferry.

On top of that, the Pentagon added its own requirements. The military needed a shuttle able to change course in the atmosphere on its return, allowing it to land at sites up to 1,000 miles to either side of its normal reentry path. This would give the Pentagon flexibility to return to Earth at will, rather than waiting several orbits until the desired landing strip came into range.

The military's "cross-range" capability required the shuttle to have some flying ability, so designers added large wings with sharp edges. The sharp wing edges cutting through the air on reentry built up tremendous concentrated heat—and that required a system of special tiles that has been a NASA nightmare.

The nightmare started with a design using 34,000 tiles on each orbiter. Each tile is unique, individually designed with its own serial number, for a specific spot on the orbiter's belly and wings. Applying a single tile can take two people half an hour, and it may require more than 200 inspection stamps. With this much work, the tiles ought to be a rock-solid defense against reentry heat. In reality, the tiles are so delicate that tiny amounts of water or a fingernail scratch can ruin them.

This fragility makes launch decisions extremely nerve-racking. If a shuttle pierces any rain clouds on the way up, the water may loosen so many tiles that it couldn't survive the trip back down. A shuttle being ferried to Kennedy atop a Boeing 747 in July 1985 accidentally went through a 20-second shower—and more than 1,000 tiles were damaged, requiring months to repair. This history makes the decision to launch *Challenger* through a hail of falling ice simply incredible.

Had the shuttle been designed solely for civilian missions, it could have used a different design. A more rounded "lifting body" would have spread the heat of reentry evenly around its bottom. Exotic tiles wouldn't be needed at the lower temperatures, avoiding the expense and delays they brought. (The tiles also added a tremendous 25,000-pound weight penalty, almost half the orbiter's payload capacity.) But the lifting body design couldn't maneuver; it required a straight-on approach to landing sites. The military vetoed it.

The shuttle was crippled in other ways. President Richard Nixon's budget watchers decided that a completely reusable shuttle would need too much up-front money, even if it might be cheaper to operate in the long run. NASA's completely reusable approach was a two-stage vehicle with both stages manned. The first stage would boost the orbiter partway to space and then fly back to the launchpad.

NASA, eager to start on a major project, settled for a partly reusable system. Two solid fuel rockets are strapped to the side of the orbiter for the initial push. Once exhausted, they parachute into the Atlantic Ocean off Florida and are dragged to shore by recovery ships; then they're broken into segments, loaded on railcars, sent to northern Utah for refurbishment and new fuel, loaded back on railcars for the trip to Kennedy, and then stacked up for another flight. This makes them "reusable" in a technical sense. But the enormous amount of work required means the actual savings are questionable. The true savings from recycling boosters, if there *are* any savings, will depend on how fast they wear out, and NASA has flown each booster only a few times.

In addition, there are safety problems with this booster scheme, as the *Challenger* tragically demonstrated. Building boosters in segments, as NASA required, is unusual for solid rockets. Most are built in one piece, and propellant is poured into them as one solid block—there are no joints to fail. NASA's segmented approach is inherently more risky because it builds in weak points at the joints.

The segments are exposed to extremely rough conditions. The boosters slam into the ocean and then undergo a 3,000-mile round-trip to Utah. These traumas often bend them out of their original round shape. These slightly squashed segments don't join properly when recycled with segments that are still circular. Mismatched segments constituted one of the causes of the *Challenger* disaster — they made it even harder for the infamous O-rings to seal the joint between segments properly.

This awkward and dangerous plan for the boosters flowed naturally from the need to portray the shuttle as a reusable spaceplane. NASA was selling the shuttle to Congress and the public on the basis of economical and reusable parts. Everything that could be recycled, therefore, would be recycled regardless of the extra risks or enormous efforts required for reusable parts. Bureaucrats and engineers all believed that intensive use — a 24-flight schedule — would justify the design.

A DISASTER WAITING TO STRIKE

The *Challenger* tragedy wasn't a fluke. Troubles started with the original contract award to Thiokol and continued to snowball. When NASA technical experts examined the bids for the booster contract, they ranked the Thiokol design dead last on design quality. But NASA administrator James Fletcher awarded the contract to Thiokol anyway, saying the company's management plan and low-cost Utah labor would keep down the space agency's costs.

Fletcher was enthusiastic about Utah's advantages — he was on the board of Pro-Utah, a lobbying group of Utah and Mormon officials, just prior to taking the helm of NASA in 1971. His decision favoring the last-place Thiokol design gave Utah a contract initially valued at $800 million. It has since risen to nearly $2 billion.

By 1978, the first tests on the Thiokol boosters showed alarming results. Company engineers had expected the pressure of ignition to force the joints between segments closed. Instead, the tests showed the joints were bent open. John Miller, a NASA manager in charge of the solid rocket motors, wrote his boss in January 1978 that the joint had to be redesigned and, in the meantime, custom-tailored shims had to be installed to keep the correct joint spacing. "Proper shim sizing and high quality O-rings are mandatory to prevent hot gas leaks and resulting catastrophic failure," his memo said. Miller's boss, Robert Eudy, did not order Thiokol to create a new joint.

(Eudy told investigators after the *Challenger* accident that waiting for a new design could have shut down booster production for two years.)

Miller tried again in January 1979. He wrote Eudy that Thiokol's position on the joint was "completely unacceptable." The way the joint bent open under pressure forced the primary seal to operate in ways that violated government and industry standards, and the secondary O-ring had been shown to be completely ineffective.

Eudy still took no action. Thiokol kept producing boosters with a joint that behaved oppositely from the engineers' original predictions.

The second shuttle flight, in November 1981, had severe trouble with one O-ring. Rocket exhaust burned through almost 20 percent of the ring's diameter. Later flights had problems as well, and by 1984 shuttle managers were talking about the level of "acceptable erosion" that would be permitted on O-rings. Their decision: Almost a third of the O-ring could be eaten away by rocket exhaust, and there would officially be "no problem."

By 1985, Thiokol engineers in charge of the joint were urgently trying to get serious support from company executives. Roger Boisjoly sent a memo to Robert Lund, the vice president of engineering at Thiokol, warning of a catastrophe: "It is my honest and very real fear that if we do not take immediate action. . . . then we stand in jeopardy of losing a flight along with all the launch pad facilities."

The record is very clear: NASA wanted a cheap booster, Thiokol executives were determined to deliver a cheap booster, and engineers who worried about safety problems just didn't understand NASA's priorities. At the time of the *Challenger* disaster, Thiokol was eligible for a $75 million NASA bonus for keeping costs under control. The bonus and future booster contracts would have been in severe danger if Thiokol executives had admitted their joint simply didn't work according to design.

NASA couldn't afford a major problem either. The U.S. Air Force was lobbying Congress for its own independent fleet of expendable launchers. The Commerce Department was trying to get its weather satellites onto expendable boosters, too.

NASA already was losing many cargoes to a European-built expendable Ariane rocket. Pushing ahead despite the warnings allowed NASA to postpone the unmasking of its shuttle as an economic failure.

MORE THAN JUST A BOOSTER PROBLEM

A congressional probe of the *Challenger* disaster found that boosters weren't the only threats to safety. The House Science and Technology Committee discovered these items:

—NASA routinely would rip apart orbiters to find spare parts for other orbiters about to be launched. Some 45 out of 300 parts needed to ready the *Challenger* had to be cannibalized from other vehicles. The danger of constantly taking orbiters apart and putting them back together is obvious.

—The main engines in the orbiter haven't had enough testing. NASA doesn't know their breaking point. NASA assumes an engine part on an orbiter is still "safe" if the corresponding part on a test engine has operated for twice as many hours without failing. The problem is twofold. First, there aren't enough test engines to establish a valid record on parts lifetime. Second, NASA frequently violates the "two times" rule when a part threatens to delay a flight.

—The orbiter's landing gear, tires, wheels, brakes and steering all are experimental. The brakes are extremely fragile. The House report says astronauts are given "the astounding instructions to never release a brake once applied because if the brake is reapplied the loose fragments will destroy it. . . . As a consequence, orbiter landings appear high risk even under ideal conditions, which seldom occur."

—Shuttle managers were attempting to track 748 problems rated "Criticality 1," meaning they were parts that had no backup. Failure of the item would destroy the mission. A system with 748 critical problems shouldn't be labeled "operational," as the shuttle was.

The Rogers Commission, the presidential panel that investigated Challenger, also found serious shuttle woes:

—Astronauts train on two simulators that are wearing out and overworked. The simulators no longer match the real world of the shuttle. Astronaut Henry Hartsfield told the commission that the simulation of the main engines was so bad it constituted "negative training."

—Each flight requires a unique batch of software to manage the mission. The shuttle has no reserve power to take care of variations from flight to flight; any change must be carefully modeled and accounted for in the software controlling the mission. The problem is that shifting payloads and the work load of nine flights in 1985 put the software team behind schedule. Often the actual flight software wasn't delivered to the astronauts until one or two weeks before lift-off, making training very rushed.

THE SHUTTLE IS A BLIND ALLEY

NASA clearly failed to create a low-cost, quick turnaround vehicle when it built the shuttle. The nation's $21 billion investment instead resulted in a spaceship requiring enormous subsidies for its continued operation. Being a drain on taxpayers is bad enough, but what's worse is that high construction and operating costs mean the shuttle can never be a leaping-off point for private space transportation. The shuttle is a blind alley for commercial space flight because no corporation can afford to operate it.

The ten-year shuttle development program also put U.S. space efforts on hold; no other initiatives could find support. NASA postponed plans for a space station, and private companies stayed away from the launcher business. The result is that space programs now lag behind the technology available in the private sector. The shuttle, after all, was designed back when engineers were still using slide rules.

The suffocating effect of NASA's dominance can be measured by comparing its progress to those of others. One comparison can be made with the Soviet program, and another can be made with the development of the airplane.

The Soviet Union Creates an Orbital Outpost

The Soviets, after their humiliating loss of the Moon race, have kept on launching payloads to low Earth orbit and the planets. And the Soviets are putting payloads into orbit at a far faster clip than the United States. In 1985 – the last year when U.S. rockets were operating normally – the Soviet bureaucracy launched 119 successful payloads compared with 28 for the NASA and Pentagon bureaucracies.

The Soviets have displayed far more vigor in launching space stations, too. The world's first space station was the Soviet *Salyut 1*, placed in orbit in 1971. The next Soviet space stations failed, but they were followed by three successful space stations in 1974 and 1975. The NASA team put up a temporary space station, the 1973 *Skylab*, using a Saturn rocket from the Apollo program. Then NASA shut down the Saturn production line and started designing the space shuttle. Now, the first permanent American space station isn't expected until 1996 at the earliest.

The Soviets continued working on space stations, launching a second-generation model in 1977. The *Salyut 6* had two docking ports, allowing unmanned cargo ships to resupply the station even while a crew ship was docked. A Soviet crew on *Salyut 7* set a 211-day endurance record in 1982, at about the same time NASA officials were still doing test flights on the space shuttle.

Early in 1986, the Soviets launched a third-generation space station called *Mir*, the Russian word for "peace." It's more automated and larger than *Salyut 7*. Soviet space crews jetted back and forth between *Mir* and *Salyut 7* in May 1986.

The fourth-generation Soviet station is now being developed on the ground for launch in the 1990s. It will house a dozen or more cosmonauts and test long-duration crew stays needed for manned flights to Mars. The Soviets already have flown a massive new booster able to lift the large-scale station sections into orbit. The new Energia booster also can carry a Soviet shuttle. The fourth-generation station, massive new booster, and shuttle will keep the Soviets ahead of NASA even after the space agency gets its own smaller space station into orbit in 1996.

The results speak for themselves: Soviet bureaucrats have made more solid progress than the American bureaucrats. The Soviets have a functioning space transportation system and two working space stations. The United States has a few very expensive and very fragile shuttles with many lingering safety problems, and only tentative plans for a space station sometime in the late-1990s.

What the Wright Brothers Mean to Space Flight

Another yardstick to measure American space progress comes from commercial aviation. How quickly did private airlines develop after the first manned aircraft mission?

Orville and Wilbur Wright started the airplane industry in 1903 with the first powered airplane flight. Just 16 years later in 1919, the Aircraft Transport & Travel Company began commercial passenger flights between London and Paris. And in 1928—just 25 years after the Wright brothers flew at Kitty Hawk—Boeing Air Transport introduced the Model 80, which carried 12 passengers and the world's first airline stewardess.

The space flight "industry" is coming along much more slowly.

Yuri Gagarin and Alan Shepard made the initial Soviet and American manned space flights in 1961. Using the airline industry as a guide, the first passenger service should have started 16 years later in 1977. Boeing should be selling a commercial spaceliner right now, more than 25 years after the

Figure 1.1
Space Development Has Lagged Historic Trends

Aircraft   vs.   Spacecraft

(Competitive market)	(Government monopoly)
Year 1 Wright Brothers fly the first powered manned airplane 1903	Yuri Gagarin makes the first orbital flight 1961
Year 16 Aircraft Transport & Travel Co. starts passenger service linking London and Paris 1919	U.S. flights grounded for six years between Apollo and Space Shuttle programs 1975-1981
Year 25 Boeing introduces the 12-passenger Model 80 1928	Challenger shuttle explodes, halting entire civilian space effort 1986
Year 27 World's first airline stewardess debuts 1930	Space Shuttle grounded for close to three years 1988
Year 28 Britain's Imperial Airways routinely flies 38-passenger aircraft throughout Europe 1931	Space Shuttle flying again — carries only a handful of government employees 1989

> **Lesson:** **Government-run space programs have failed. The U.S. should return to private initiative spurred by government incentives.**

start of manned space flight. Boeing is not even *considering* a passenger space vehicle developed with its own money.

Thus, we are far behind the pace set by Soviet space bureaucrats—and far behind the pace set in the development of commercial aviation. The cause is our unprecedented reliance on government officials to lead the exploration of a lucrative new territory and technology. NASA has proved that federal bureaucrats aren't able to provide affordable space transportation.

Current policy prohibits NASA from launching commercial satellites. The shuttle fleet is reserved for military payloads, planetary probes, microgravity research, and a few foreign and U.S. government satellites. Because NASA and many members of Congress aren't happy about losing commercial satellite payloads, the official policy is under fire.

If the noncommercial launch policy survives, private operators can repeat transportation history. Private railroad companies opened the American West, spurred on by land subsidies. The airline industry flourished under private control despite being the 1930s equivalent of extremely high technology. All the government provided was guaranteed airmail contracts, not a federal aviation authority flying all the planes. We can achieve progress on the space frontier, too, if we give up government control and rely on government incentives.

Transportation costs always drop as a frontier develops. Competitors fight to win markets, and companies specialize in hauling different cargoes: people, building supplies, food, or whatever. The orbital frontier is desperately overdue for the benefits of competition and specialization. The private sector must push NASA out of the way for the next stage to begin.

The Birth of a New Industry 2

NASA's space shuttle turned out to be a blind alley of high costs. Now we are retracing our steps, back to the throwaway rockets used before the shuttle.

Several aerospace contractors built expendable launch vehicles (ELVs) for NASA in the 1960s, and they have restarted production. These government-surplus designs are now serving their first commercial customers. They cost less than the shuttle — but not much.

At least four small firms are starting from scratch on all-new industrial rockets. Some are trying to offer lower prices than the shuttle and recycled 1960s rockets. Others hope that small rockets tailored for small payloads will find a market niche, even if the cost per pound is more than the shuttle.

The resurgence of expendable boosters will not make space the next trendy vacation spot. Their costs are still too high. But they will start a historic shift to free enterprise in space.

FRESH DESIGNS TO CUT COSTS

The firms starting with a fresh slate have tremendous advantages. New rockets will be the first to use the electronics revolution that turned big vacuum-tube TVs into tiny Sony Watchmans. Cost and reliability will vastly improve.

Even the basic construction materials are new. Instead of metal tanks and walls, the new rockets can use lightweight composites. These composites are superstrong sandwiches of a plastic and a reinforcing fiber. Besides being lighter than metal, composites are easier to shape into circular rocket walls than metal. Low-cost automated manufacturing equipment can mass-produce composite structures, giving them the advantage in both weight and price.

The new rockets *won't* have revolutionary engines. Their engines may even be less powerful than existing designs. By taking a step back from race car performance, companies can manufacture rocket engines at lower cost. Parts don't require exotic materials and hair-thin tolerances if the engine runs at only 80 or 90 percent of the maximum possible temperature and pressure.

For example, nozzles at the end of today's rockets are very expensive. They are handcrafted from layers and layers of costly material to resist the exhaust flame while weighing as little as possible. Nozzles for new rockets might not try to resist the flame. Instead, they could be designed to melt or burn away gradually. One approach is to make them from a slow-burning propellant poured into a nozzle-shaped mold. The nozzle material would burn off as the rocket rises. The nozzle would be heavier than current designs, but in return it could be mass-produced from extremely cheap materials.

The new rockets will use many engines instead of just a few giant ones. Private companies favor clusters of several small engines because they can be mass-produced. Testing small engines is easier, too. Dozens can be put through destruction testing, firing until they fail. The huge space shuttle boosters need this exhaustive testing, but each one is too expensive to use on deliberate failures. An executive of Morton Thiokol told investigators he didn't think that "there's enough money in the national treasury to do that."

Private operators will seek engines, structures, and electronics that strike a balance between performance and expense. They will reject the "best" solution if it costs too much. And they will chose a simple system over a complicated one, even if the complex one delivers a little more thrust. The new rockets will be the Toyotas of space, instead of the Porsches designed earlier.

A BRAND-NEW ROCKET FOR $10 MILLION

Many people doubt corporations can afford the multibillion-dollar costs of creating new launchers. It's true that only NASA can afford to create systems using NASA methods. But private operators assembling existing devices in fresh ways can create a new breed of rockets.

The story of a small Redwood City, California, company proves the point. Starstruck Inc. successfully launched a suborbital rocket of its own design in 1984. Michael Scott, a former Apple Computer president, supplied most of the $10 million Starstruck needed.

The Starstruck vehicle combined the best features of liquid and solid rockets to create a hybrid. In a standard solid rocket, the fuel and oxidizer are mixed together and cast into a mold. This creates a rubbery cylinder with a hollow core. When ignited, the motor fires until the entire cylinder burns out. In a liquid rocket, the fuel and the oxidizer are held in separate tanks. Controlling the flow of fuel and oxidizer into the engine varies the thrust level.

The Starstruck engine was a combination of the two types. Only the fuel was in solid form, cast into a hollow cylinder. The oxidizer was liquid – a tank of liquid oxygen sprayed into the hollow core of fuel for combustion. The engine could be stopped and started by the flow of liquid oxygen. The thrust control made it more flexible than solid rockets that keep firing until they're exhausted.

The Starstruck hybrid engine also had better safety credentials than either solid or liquid rockets. Solid rockets are extremely hazardous to manufacture because the propellants can explode while being cast into cylinders. Liquid rockets are dangerous on the pad and during launch, because the fuel and oxygen tanks can rupture and turn the vehicle into a fireball.

The Starstruck design has neither problem. During manufacturing, only the fuel is cast into the hollow cylinder. Fires are possible but not explosions. Risks on the launchpad also are lower. If the tank of liquid oxygen ruptures during launch, it can't mix explosively with the entire mass of fuel; only the surface of the solid fuel is available to burn.

To improve safety and cut costs further, Starstruck launched its rocket from water. The company loaded the rocket on a barge and sailed out into the Pacific Ocean. About 200 miles off shore, engineers dumped the rocket overboard. Weights on the bottom forced it to point skyward until takeoff. The water launch completely isolated the rocket from any innocent bystanders and saved money by eliminating the need for a launchpad and launch tower.

In practice, the water launch had horrible results. Small equipment problems turned into big delays because technicians couldn't reach the rocket to fix them after it hit the water. Even before the rocket went overboard, technicians were leery of making repairs while on a pitching and rolling barge. Starstruck had three false starts before its final success, largely because the water launch made last-minute repairs impossible. The final attempt was made only 50 miles out, in the shelter of St. Nicholas Island, partly to overcome problems with waves that had plagued earlier attempts.

Established rocket engineers criticized the Starstruck hybrid design as inefficient. Some solid fuel would remain unburned because the liquid oxygen couldn't be sprayed evenly over the entire surface of the fuel.

The criticism hits the central debate in reducing space costs. Up to now, space engineers have always fought to improve the theoretical efficiency of engines and spacecraft. The costs—in design and operation of the high-performance craft—were secondary. Engineers tend to be more interested in reaching the highest technology, not the lowest costs.

The new breed of space entrepreneurs temper their enthusiasm for theoretical efficiency with more respect for operational costs. Propellants only amount to 1 percent of the shuttle's launch costs, for example. Fuel for the Titan amounts to only 2 percent of its launch costs. So if a small fraction of the fuel is wasted in a hybrid engine, that's only a tiny sliver of the overall expense.

The hybrid engine's real-life efficiency shows in lower manufacturing and support costs. Starstruck poured the fuel into molds in a neighborhood industrial park. That's impossible with solid rocket engines that mix the fuel and oxidizer together in liquid form before casting. The mixing must be done in isolated, heavily shielded buildings—manufacturing plants tend to blow up occasionally.

The hybrid's operational advantage over engines using liquid hydrogen and liquid oxygen comes from avoiding hydrogen. It uses only liquid oxygen, which is the easier one to handle. Liquid oxygen is a standard industrial chemical that can be stored in a simple insulated tank, like a giant Thermos bottle. No active refrigeration equipment is needed, so liquid oxygen is cheap to keep around. The oxygen supplier does the hard part, getting it chilled down to -297 degrees Fahrenheit. After delivery, as the oxygen gradually absorbs heat, it simply boils away slowly. Liquid oxygen automatically stays at -297 until the last drop boils off, just as water hits a maximum temperature of 212 degrees and gets no hotter.

The liquid hydrogen used with the shuttle and other rockets is another story. Liquid hydrogen is an exotic chemical with few industrial uses. It must be kept much colder than liquid oxygen, at -423 degrees. When air hits the extremely cold sides of a liquid hydrogen tank, it condenses into a liquid and often starts to cause trouble. Liquid air will drip into surrounding equipment and freeze it solid.

The Starstruck engine got some of the benefits of liquid fuel—the ability to stop, start, and throttle—without the burden of handling liquid hydrogen.

The $10 million Starstruck experience began in March 1981 and ended in August 1984 after three rockets were built and one launched. (Two remain in a San Jose warehouse.) The effort led only to a suborbital rocket, sometimes called a sounding rocket. Sounding rockets aren't nearly as versatile as larger rockets that reach orbit, but they do have a market. NASA launches several dozen annually to study the upper atmosphere, cosmic rays, short-term weightlessness, and other subjects.

To reach its eventual goal of putting communications satellites into orbit, Starstruck needed to spend additional millions to develop a larger rocket. This new round of financing was part of the founders' original plan but never took place. Scott shocked company investors and workers by closing the firm ten days after the demonstration flight.

James Bennett was one of the original Starstruck executives frustrated by the abrupt shutdown of the company. He speculates on why Scott apparently never bothered to seek the additional financing needed to continue operations: Scott already disliked sharing authority with Starstruck's original founders, and bringing in new financial backers would have diluted his control further.

After closing Starstruck, Scott immediately started a rocket firm called White Rose. He hired ten key Starstruck employees and operated in Starstruck's offices for several days before shutting down entirely.

THE AMERICAN ROCKET COMPANY

Several veterans of the Starstruck adventure gathered in May 1985 to found a new firm. The American Rocket Company (AMROC) in Camarillo, California, is again working on a hybrid design. The Starstruck originals include Bennett and engineer Bevin McKinney.

Joining them are executives with substantial business experience. AMROC's chairman, Stuart Kreiger, was a cofounder of Planning Research Corporation, a firm with sales approaching $500 million in 1986. Other senior AMROC executives held high-level engineering and management jobs at major aerospace corporations and the Jet Propulsion Laboratory.

AMROC's first rockets are being designed to boost satellites weighing one or two tons into low Earth orbit, about 150 to 300 miles high. The launch fee will be in the range of $8 million, or a cost per pound of perhaps only $2,000. That's far lower than the $6,000 per pound cost of the shuttle and about 50 percent less than the price of government-surplus ELVs.

The company is using the Rocket Propulsion Laboratory at Edwards Air Force Base for engine tests. AMROC refurbished an unused test stand and has successfully fired several engines.

Can AMROC Hit Its Target?

AMROC is a small company with several dozen employees, aiming to break cost barriers that have defied NASA and its massive contractors. Its executives give several reasons why it will succeed at truly low-cost freight delivery to space:

— They are designing an industrial rocket with off-the-shelf parts. Components will not be "MilSpec," which means built to military specifications that require careful documentation of each part and each manufacturing step. AMROC will also slash the number of parts used, reducing both assembly costs and points where failures could occur. For example, there are no pumps to move the liquid oxygen. Instead, gas forced into the oxygen tank prior to launch will squirt it into the combustion chamber.

— They are operating as an entrepreneurial start-up company instead of as a defense contractor. Shop floor workers talk directly to the rocket's top engineers when they hit construction problems, and design changes can be approved in a few hours. AMROC executives have worked in aerospace companies where a simple change from a $12 valve to a $17 valve took six to ten weeks to accomplish. One chain of command had to be convinced the new valve was needed, another chain of command then had to approve the extra cost, and another set of officials had to issue the actual contract documents authorizing the design and cost changes.

— The AMROC vehicle builds on two decades of computer and electronic advances that are missing from boosters designed in the 1960s. Electronics are a big chunk of any rocket's costs, far larger than the cost of the fuel. Company engineers say today's computers are so compact they can put any electronics they want in the rocket without worrying about the added weight. Reliability is better, too—a rocket can carry several backup systems because the extra weight is trivial.

— Powerful engineering terminals make design tremendously efficient. "We do with a single nerd and personal computer what it takes the Air Force ten people [to do]," AMROC president George Koopman says.

Overstaffing is a prime cause of high launch costs, Koopman says. He notes 600 people are required to launch a Delta rocket under NASA's control, but only 60 are needed when the Air Force does it. That astounds Koopman because the military is notorious for poor use of manpower.

The revolutionary impact of personal computers stands out when AMROC conducts engine tests. Company technicians monitor all the important valves, tanks, and switches on a Macintosh. The entire engine appears on the screen as a sketch of fuel lines and components — like adding animation to an engineer's drawing. Once the operators see that all parts are ready for ignition, a simple click of the mouse fires the engine.

Substituting a $2,000 personal computer for the million dollar mainframes used in the 1960s is just one way a small company can hope to succeed in today's space age. Another method is to keep initial investments small, which is why AMROC chose to create an expendable rocket (Figure 2.1). Reusable rockets or spaceplanes would be far costlier to design, as engineers struggle to eliminate every excess pound — a low-weight design is crucial to bringing down long-run operating costs. But the battle to reduce weight also tends to reduce strength. The flimsier the design gets, the more key parts are likely to fail. These conflicting goals — light weight to reduce costs versus strength to keep the rocket operating over a long lifetime — make development of reusable vehicles time-consuming and expensive.

High development costs put a tremendous economic burden on reusable spacecraft. To make the big investment pay off, they must be used intensively over a long period. That's the rule that drove shuttle managers to seek every possible launch customer. The second-generation shuttles of the late 1990s will face the same question: Can they find enough customers to pay back the investment in a reusable rocket or spaceplane? AMROC couldn't take the risk and opted

Figure 2.1
AMROC's ILV-1

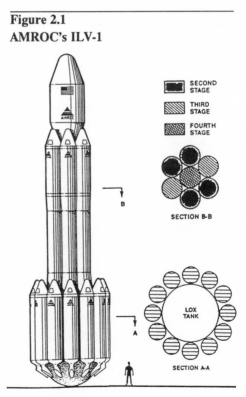

Source: Drawing courtesy of the American Rocket Co.

for an expendable rocket.

AMROC executives cite a classic argument that others use to justify reusable spaceplanes: The argument asks, Would it make sense to build a Boeing 747, fly it once across the Atlantic and then junk it? Obviously not. But instead of one flight, suppose the plane only flies 50 times during its entire life. The result is almost as silly—the $125 million plane would cost $2.5 million per flight or $7,000 per passenger ticket.

That's the situation facing NASA. Each of its shuttles may fly only 50 times before being overtaken by new technologies. Reusable vehicles must fly intensively to pay back development and construction costs. Throwaway boosters may be cheaper for short-term or specialized applications.

The initial AMROC vehicle will be able to reach only a very low orbit, the same one traveled by the space shuttle. This low altitude won't work for standard communications satellites. They aim for a geostationary orbit of 22,300 miles up, where their forward movement is exactly matched by the turning of the Earth. Communications satellites launched by the shuttle use an additional rocket to send them up to geostationary orbit. AMROC's initial booster won't be powerful enough to launch both a satellite and this upper stage. AMROC must seek out customers who are satisfied with a much lower orbit.

Low Earth orbit (LEO) is great for remote sensing of the Earth with photography or radar. For example, a Japanese company might buy a LEO remote sensing satellite to track ocean currents and temperatures. Then it can flash private advice to its fleet of trawlers on the best regions for commercial fishing. A U.S. grain trader such as Cargill might buy a satellite to create a private forecast of worldwide crop yields. Mining and oil companies could buy remote sensing satellites to spot the best targets for oil, coal, and mineral exploration.

Even communications is possible through LEO satellites. They can't handle telephone calls because they move out of range after a few minutes, but written or voice-mail service is possible. For example, a LEO satellite can collect and distribute electronic mail on every pass it makes above a region. Amateur radio buffs already have lofted several experimental satellites that send electronic mail through this kind of "store-and-forward" function. Oscar 11, for example, flies a polar orbit that covers the globe twice a day. A fleet of such satellites could provide connections every hour or even every few minutes.

These electronic mail satellites can serve smaller and less expensive ground stations than today's geosynchronous satellites. The lower orbit means less power is needed to punch a signal through to the satellite. The lower cost could make satellite links a standard accessory for personal computers. Executives and researchers in Third World countries would especially treasure a direct link to the advanced nations, because phone and telex service can be very expensive and unreliable. Direct satellite links behind the iron curtain would be fascinating—authorities might find them even harder to control than their current plague of copied video cassettes.

To entice customers who aren't experienced satellite builders, AMROC is offering a package deal: launch plus a satellite shell already built to handle routine communications, control, and electrical supply. All the customer has to provide is the actual experimental package. The shell can hold from 12 to 16 payloads. The package deal price for one of the 50-pound payloads is under $1 million. Called Orbital Express, the service is a joint venture of AMROC and Globesat Inc. of Logan, Utah. It received several hundred inquiries from potential customers immediately after the package deal was announced in summer 1987.

The military is another customer for LEO launches. The Strategic Defense Initiative will need cheap satellites for target practice, and some spy satellites work best in low Earth orbit as well. Navy applications include antisubmarine warfare and over-the-horizon targeting.

The Defense Advanced Research Projects Agency (DARPA) is the leading military enthusiast for small-scale satellites. (They've been nicknamed "lightsats" and "cheapsats.") DARPA proposed a lightsat program in 1986 costing $100 million per year, but that scheme was shot down in 1988 by former Air Force Secretary Edward Aldridge. It later was reborn, with funding in the $30 million per year range, after managers agreed to concentrate on small satellites rather than small launchers.

AMROC almost collapsed after the October 1987 stock market crash made investors more cautious. But it managed to stay in operation and even attract solid corporate attention. In 1988, it teamed up with a subsidiary of Westinghouse Electric Corporation to jointly study ways of reducing launch costs.

ORBITAL SCIENCES CORPORATION

Unlike AMROC, several private rocket companies aren't promising low costs per pound. Instead, they hope small rockets can find small jobs that aren't suited for the big boosters.

Orbital Sciences Corporation (OSC) began plotting its entry into the lightsat market in 1987. The suburban Washington, D.C., firm formed a secret team to hatch a small rocket that could be launched from under the wing of a B-52. Joining with OSC to create the Pegasus rocket was the Hercules Corporation, a maker of solid rocket motors.

The Pegasus project became public in 1988 when OSC won a DARPA contract for $6.3 million. The cost per pound was $10,000, triple the cost of a large Titan or Atlas booster. OSC hopes Pegasus can win customers on convenience rather than per-pound efficiency.

The greatest convenience is the ability to launch a small satellite on demand, without waiting for spare room to become available on the shuttle or a large booster. The shuttle and large boosters also may not be heading to where the small satellite wants to go. Satellites can be aimed for low Earth orbit, anywhere from 100 to 300 miles high, or geosynchronous orbit, 22,300 miles up. Some navigation satellites even get sent to an in-between orbit 11,000 miles high. The inclination of the orbit is another variable. A zero-inclination orbit is one directly above the equator, the route traveled by geosynchronous communications satellites. A polar orbit, which remote sensing and spy satellites often use, is a 90-degree inclination. The shuttle usually flies at 28 degrees.

So for the small satellite needing a particular orbit and inclination, waiting to hitch a ride on just the right shuttle or large booster rocket can be chancy. The paperwork can be daunting as well—NASA or the booster's main customer always wants absolute proof that the hitchhiker satellite won't endanger the main mission. Buying your own small rocket cuts through all the problems, even if the per-pound cost is higher.

For the military, the B-52 launch is appealing because it allows greater secrecy. Standard ground launch sites are all under surveillance by Soviet intelligence. Pegasus can be launched from anywhere in the world, to strike targets anywhere in the world. For reconnaissance missions, the B-52 can position Pegasus so that it gets pictures of the target on the very first orbit. That's often not possible with ground sites, which for safety reasons can't send an ascending rocket over inhabited regions. For example, the shuttle can't be launched into a polar orbit from Kennedy Space Center because it would pass over the mainland United States.

OSC was founded in the early 1980s to complement the services offered by the shuttle to communications satellites. The shuttle only reaches low Earth orbit, whereas most communications satellites need to reach the far higher geosynchronous orbit. OSC planned to supply those satellites with the "upper stages" they need to bridge the gap.

The *Challenger* disaster spelled the end to that business, as most commercial satellites were ordered off the shuttle. The resourceful OSC executives still managed to find some customers for their Transfer Orbit Stage (TOS). They sold one to NASA for a shuttle-launched Mars probe, for example. But they clearly needed a new line of business, and their lightsat initiative seems likely to succeed.

SPACE SERVICES INC.

Another rocket company offering convenience rather than low price is Space Services Inc. (SSI) of Houston. SSI gained fame in September 1982 when it launched a suborbital Conestoga I rocket. The Conestoga was put together by a subcontractor, Space Vector Corp. of Northridge, California. Space Vector used a Minuteman rocket engine "leased" from the military — so neither Space Services nor Space Vector actually broke new ground in constructing rocket engines.

Space Vector and SSI parted company in 1984. SSI continues to use the same strategy, however, of assembling existing rocket engines into new configurations. SSI believes that buying off-the-shelf rocket engines will save it money compared with AMROC's home-built approach.

SSI won another round of publicity in January 1985 when it signed its first customers — funeral directors who wanted to loft cremated remains into a celestial resting place. Four years later, SSI hadn't made any announced progress on putting ashes into orbit.

In between the Conestoga I launch and the orbiting mausoleum, SSI tried to create a remote sensing business. The remote sensing operation was to build custom satellites and analysis software for private users and thus create a market for SSI's rockets. This effort also failed to create customers.

SSI's latest aim is to participate in military programs, one of the markets also sought by AMROC. SSI is asking $15 million to $20 million per launch, twice the price AMROC hopes to charge. The company won a $300,000 DARPA study contract in 1988.

Another new market is launching suborbital rockets that provide a few minutes of microgravity for research projects. NASA has funded several university-based consortia that need this type of service. SSI became the first launch firm to serve this market in March 1989, when it launched the Consort I under a $1 million contract from a consortium based at the University of Alabama-Huntsville.

The Consort I used a Morton-Thiokol engine for the first stage and a Black Brandt VC motor made by Bristol Aerospace of Canada for the second stage. The rocket reached an altitude of 190 miles, eight short of the planned height, and landed 58 miles downrange from the White Sands, N.M. missile range.

Real estate magnate David Hannah, Jr., has provided most of the cash for SSI. Former astronaut Deke Slayton provided the name recognition. Both struggled for years to provide a sound business plan that would get SSI into full operation. They got one boost in 1987 when the local power company, Houston Industries Inc., bought an option to take control of SSI, and another when the firm began suborbital flights in 1989. If the utility follows through and takes control, SSI would receive the major funding needed to launch orbital demonstration flights.

PACIFIC AMERICAN LAUNCH SYSTEMS

Rounding out the quartet of active new companies is Pacific American Launch Systems, located in Menlo Park, California. Pacific American is run by Gary Hudson, a space visionary who designed Space Service's unsuccessful first rocket, the Percheron, in 1981.

Hudson later drew up plans for a passenger rocket called the Phoenix — which Society Expeditions started touting in September 1985 as the basis for its widely advertised space tourism package. The $52,000 ticket offered by Society Expeditions drew 300 customers, each willing to put down a $5,000 deposit. The company's ability to fill a half-dozen flights with only a paper design for a spacecraft shows the public hunger for ways to participate personally in space exploration.

Society Expeditions and Hudson later parted ways. Society Expeditions eventually dropped out of the space tourism business.

In 1986, Pacific American proposed a new rocket, the expendable Liberty 400. The rocket won an extremely important endorsement in 1988 from the Army Strategic Defense Command. The Army paid $402,000 for a proof-of-concept test — a demonstration firing of the first stage at Edwards Air Force Base.

The Liberty rocket aims for a $2 million to $4 million launch price, to put 300 to 500 pounds into low Earth orbit. If Hudson can reach the lower end of his goal, the price would undercut both OSC and SSI.

Hudson's rocket would use liquid fuels, like the shuttle main engines, but they'd be far simpler and cheaper. The shuttle main engines use complex turbopumps to force fuel into the combustion chamber. The Liberty engines rely on gas pressure instead. Helium will be forced into the fuel tanks, and that will squirt the fuel out.

Pressure-fed motors aren't as powerful as pump-fed motors, but they are far less complex. Hudson's gamble with the Liberty hinges on whether basic low-tech designs can slash costs enough to offset the lower performance.

Pacific American had only seven employees at the time of the Army contract, a midget compared to OSC and AMROC. (And they are midgets by comparison to the makers of the large 1960s boosters.) Hudson reported that investors had put about $1 million into the firm at the time of the Army test and would add $5 million to $10 million more if Pacific American wins an actual launch contract with the Pentagon.

Other space vehicles have been announced, but none seem to have as much backing as AMROC, OSC, SSI and Pacific American.

OLD ROCKETS PUT BACK IN SERVICE

Old-style rockets offered by major aerospace contractors were very reliable during the 1970s and early 1980s. Then they suffered a sudden rash of failures. The military's huge Titan started the chain of woe with an explosion in August 1985. It failed again after the *Challenger* disaster, as did a Delta rocket.

The Titan is the Air Force's favorite rocket. It's had steady work despite the advent of the shuttle, making it the most up-to-date of all the existing boosters. Martin Marietta, the prime contractor for the Titan series, is selling commercial launches without any government subsidy.

Martin Marietta's bold no-subsidy stance for commercial launches rests on its strong military sales. The Air Force never liked NASA's "shuttle only" policy and in 1985 lobbied for backup Titan boosters. The generals correctly argued that total reliance on the shuttle made the military too vulnerable to calamities like the *Challenger* tragedy. NASA fought bitterly against the Titan plan, delaying it for months, but the White House finally endorsed the Air Force logic.

The Air Force awarded Martin Marietta with $2.2 billion in Titan contracts. One contract calls for 23 new Titan IV boosters. The other contract involves conversion of retired Titan II strategic intercontinental ballistic missiles (ICBMs) to satellite launchers. Coupled with its ongoing launches of military communications and spy satellites on Titan IIIs, Martin Marietta has a tremendous manufacturing base for its commercial rocket line.

The Titans will be the military's main machine because the Air Force uses heavy satellites for spying, communications, attack warning, and weather forecasting. The top-of-the-line Titan IV is strong enough to carry most shuttle-sized cargoes.

In addition to military cargoes, some heavy civilian science projects could use the Titan IV. NASA would like to shift its Mars Observer mission from the shuttle to a Titan IV.

Most commercial satellites don't need the Titan IV's tremendous lifting power. Martin Marietta is offering them the smaller Titan III. Federal Express was the first commercial Titan customer to sign up, putting down a $100,000 deposit for a 1989 launch. Actual launch prices are confidential but Martin Marietta opens the bargaining by quoting a $100 million price.

THE AIR FORCE PLAYS KING MAKER

The successful Air Force campaign in 1985 for a new series of heavy Titan IV rockets spawned a second campaign in 1986. The Air Force began lobbying for a medium-duty expendable rocket series.

Four companies won contracts to design the medium expendable rocket. The specifications seemed suited for only one existing rocket—the Atlas-Centaur built by General Dynamics. The Air Force surprised the aerospace industry in 1987 by awarding the contract to McDonnell Douglas and its small Delta. The price offered by McDonnell Douglas was so attractive that the Air Force accepted the Delta's low power. The company promised to upgrade the Delta's lifting strength gradually as the Air Force creates heavier satellites.

The Air Force bought seven Deltas initially at a price of $45 million each to launch Global Positioning System (GPS) satellites. Warships, bombers, and tanks use GPS satellites to find their precise locations instantly. Headquarters uses GPS satellites to spot nuclear explosions, either Third World atmospheric tests in peacetime or the initial craters of World War III.

The Air Force plans a second batch of 13 Deltas for launches in 1990 and 1991. McDonnell Douglas will sell these for only $27 million each— total contract value: $985 million.

The rich Air Force contract propelled McDonnell Douglas into the commercial launch business almost against its will. Before winning the Air Force competition, McDonnell Douglas wanted NASA to buy ELVs and take the risk of reselling them to commercial customers. Now the military will pay for upgrading the Deltas and for all the firm's overhead in manufacturing them.

In fact, the Air Force is outspoken about wanting to help commercial rocket sales. Former Air Force Secretary Edward Aldridge, Jr., said the contract is expressly designed to encourage American competition with European, Chinese, Soviet, and Japanese rockets. McDonnell Douglas says it could produce 12 Deltas a year, leaving room for four or five commercial launches annually.

McDonnell Douglas quotes a $50 million price for commercial Delta launches. In view of the $27 million paid by the Air Force for its second batch of Deltas, there probably is room to negotiate. (The Air Force launches go to a lower orbit than most commercial satellites need, which is one reason for the lower price.)

Having the Air Force select McDonnell Douglas to receive this golden handshake isn't precisely the free market at work. But it may be better than leaving things to chance. The established aerospace contractors all supported NASA's "shuttle-only" policy, and only Martin Marietta was aggressive in spending its own cash to leap back into commercial launches.

The initial group of seven Deltas could put payloads into low Earth orbit for roughly $4,500 a pound. That's not much better than the shuttle cost, but the Air Force is absolutely desperate to get the GPS satellites into orbit as soon as possible. The second run of 13 Deltas offers a much better efficiency, roughly $2,750 per pound to low Earth orbit. (The actual GPS target is not low Earth orbit but a unique altitude halfway between low Earth orbit and the high geosynchronous orbit of communications satellites.)

EUROPE WINS CONTRACTS FOR ARIANE ROCKET

The only company with a substantial track record in launching commercial satellites is Arianespace, a European firm supported by several governments, with France in the lead. Arianespace offers a series of Ariane expendable boosters launched from South America, in French Guiana. The company is partly driven by political goals: High-tech prestige and full employment for its aerospace subcontractors are just as important as making a profit on each flight.

Before the *Challenger* disaster, Arianespace set prices low enough to beat NASA on many contested satellite contracts. After *Challenger*, potential Arianespace customers discovered prices had almost doubled. The typical satellite owner would now pay about $38 million if it shared a launch with another customer, or about $75 million if the satellite was heavy enough to require a dedicated flight. U.S. firms eager to find an alternative to the shuttle came back from France shocked at the new prices.

Arianespace says the weak dollar, not just the absence of subsidized shuttle competition, contributed to the escalation. Customers accepted the higher costs because the Ariane series is heavily booked through 1991. The order list of 40 satellites was worth more than $2.3 billion.

American firms trying to market existing NASA boosters will find Arianespace tough competition, because it can draw on the same taxpayer subsidies that fueled the shuttle. The governments behind Arianespace will make sure it wins enough launch contracts to stay busy. The official European policy is to subsidize the development of new Ariane versions but not their actual operation.

American firms also can hope that Arianespace continues to suffer regular launch failures. Four of Arianespace's first 18 launches exploded or ditched in the ocean. A third-stage failure on a May 1986 launch kept Ariane grounded for more than a year. American rockets had better records — until their chain of disasters in 1986-87 — giving them a reduced rate on crucial insurance premiums.

Customers need reliability so they can be sure their satellites will start earning money on schedule. A company forced to wait for a replacement satellite has its business plans thrown into turmoil, and that cost usually isn't covered by insurance.

The 1986 grounding led to a new attitude at Arianespace. Now after every flight, its engineers carefully sift through the rocket's performance, looking for signs of potential problems. Every deviation from the perfect flight pattern is analyzed and eradicated before the next launch. Engineers no longer congratulate themselves for simply making it into orbit without a disaster.

Communist Rockets Start Price War

The Soviet Union cranked up its sales efforts for its Proton rocket after U.S. and European failures in 1986. The Soviets had made a halfhearted attempt once before, in 1983. Their target markets are international organizations and Third World countries. One organization, Inmarsat, is the world agency for satellite links to ships and drilling platforms. The Soviets own 14 percent of Inmarsat and thus can force serious consideration of Proton launches.

The Soviets also are trying very Western marketing techniques. The space shuttle won the Arabsat contract, for example, by agreeing to fly a Saudi Arabian prince on the flight that launched Arabsat. The Soviets also are offering to fly Western astronauts, including one from the United

Kingdom. A French astronaut already has made the trip, with a second trip scheduled.

Another capitalist tactic is price cutting. The Soviets are undercutting Western prices by substantial amounts. The announced price of a Soviet launch has bounced around but seems to be about $30 million for putting 4,000 pounds into geosynchronous orbit. Special discounts are available for Third World customers.

The barrier to Proton sales is a U.S. ban on technology transfer. American bureaucrats can ban export of any U.S. technology under the Munitions Control Act. And the Paris-based Coordinating Committee for Multilateral Export Controls enforces a worldwide ban against sending high technology to communist countries. Satellites have immediate military potential, so export licenses aren't likely—especially when they take business away from domestic rocket companies.

Dmitry Poletayev, head of the Soviet State Committee for Space, says customer payloads will be "secure" during launch preparations. Poletayev also says customers can travel with their satellites to the Baikonur cosmodrome and oversee them right to the moment of launch. The catch is that the Soviets have technology to inspect the interior of satellites from a distance, so even a continuous guard over the satellite couldn't prevent the potential leak of secrets.

Satellite owners privately are very skeptical about Soviet promises to keep their hands off U.S. satellite technology. One executive noted that the Soviet Union doesn't have a single bureaucracy making decisions—it has three. The communist party is nominally in control, but both the military and the KGB security apparatus have tremendous power. In the constant jockeying for dominance among the three, any deal made with the party could easily be violated by one of the other powers.

These problems make the Proton's future in the West rather bleak. Perhaps in recognition of this, the Soviets came up with another idea: selling communications services instead of boosters. They have offered to launch Soviet communications satellites into any orbital slot requested by a customer, with that customer then leasing time on the satellite. The plan may solve the problem of export controls.

The Soviets also have offered to sell launch services on their new shuttle and Energia booster. The Energia can put 100 tons into orbit, four times the capacity of the U.S. shuttle.

The communist country actually winning Western customers is China with its Long March rockets. Their CZ-2 and CZ-3 models are the right size for the average communications satellite, yet they are priced at 20 to 30 percent below Western rockets. China is on better terms with the United States than is the Soviet Union and in 1988 the Reagan administration approved the export of two Western satellites for launch on the Long March.

The Chinese are quite earnest about marketing their boosters, sending top-ranking engineers to conferences in the U.S. and Europe. The chief Long March engineer, Huang Zuo Yi, told one audience that China's approach to rocketry is "KISS — Keep It Stupid and Simple." (The American version of KISS is "Keep It Simple, Stupid" but perhaps Huang thought that was impolite.) The Chinese KISS shows in the way engines are operated far below their design standards. The engines are built to deliver 85,000 pounds of thrust but are operated at only 71,000 pounds. And they were tested to survive 1,000 seconds of firing, even though they are used for only 150 seconds.

Huang said Western customers could expect to launch their satellites only 15 months after signing a contract.

The "KISS" of simplicity led the Chinese to develop the world's only partly wooden satellites. The People's Liberation Army's reconnaissance satellite reenters the atmosphere carrying film exposed in orbit. The heat shield on the recovery capsule is made from wood. The Chinese say a thick layer of oak burns slowly and carries away the heat of reentry.

Eight spy satellites with wooden heat shields have flown, and China is offering the basic spacecraft for foreign missions. The round-trip feature of the recovery capsule provides a way for materials science experiments to return their results to Earth for analysis.

The first customer for this round-trip service was a European consortium called INTOSPACE, which flew a 44-pound pharmaceutical experiment in August 1988. The Chinese charged less than $1 million. INTOSPACE officials watched the launch from a Gobi Desert launchpad only seven months after the project was initially approved.

Insurance is a touchy problem for Chinese launches. Western insurance companies are reluctant to underwrite the Long March launches because its long-term reliability is unknown.

Despite insurance worries, the Chinese gathered an impressive number of launch customers while the Soviets were empty-handed.

In February 1986, Sweden's Svenska Rymdaktiebolaget became the first foreign company to sign an option on a Chinese rocket. Although the Mailstar message-relay satellite was later canceled, the 1991 launch reservation will be used instead by a Swedish scientific satellite.

The most lucrative Western contracts cover satellites for the Pacific Rim. The first is to be Asiasat, the new name for the Westar 6 satellite that U.S.

shuttle astronauts rescued from a useless orbit in 1985. The satellite was sold by the insurance companies that owned it and ended up in the hands of a Hong Kong consortium of Chinese and British interests. It's scheduled for launch in 1989.

Two brand-new satellites built by Hughes Aircraft Company are to reach orbit in 1991 and 1992. They were built for Aussat to serve Australian communications.

Brazil will orbit a remote sensing satellite in 1992, and Pakistan may join the list. The Chinese are definitely striving for a world-class space program. They probably will begin launching their own astronauts in the 1990s.

THE NEXT FIVE YEARS

The commercial launch market now is dominated by Arianespace. McDonnell Douglas, Martin Marietta and General Dynamics are rather far back, fighting for second place. Unlike Arianespace, commercial launches are just a sideline for each firm. The Chinese are coming up fast to challenge for the number three or four position. AMROC and Pacific American are daring darkhorse contenders for the early 1990s but could founder at any time from lack of cash.

To insure a healthy American launch industry, some short-term government help is needed. Offering guaranteed payloads would be one possibility. The Space Incentive Act written by the Citizen's Advisory Council on National Space Policy proposes this structure:

The United States would guarantee every year to pay $500 per pound for the first 1 million pounds placed into orbit by private U.S. companies. The money would be paid only for performance – actual delivery of payloads into orbit. There'd be no progress payments, and the government would have absolutely no role in picking which companies got funding. Companies would obtain their development money from investors and banks. These private experts would make the judgments about which firms were technically competent and thus able to collect on the government guarantees.

The annual cost of the guarantees – a half billion dollars – would be tiny compared to the benefits of the revolutionary $500 per pound price. NASA spent $21 billion on the shuttle and actually increased the per-pound costs, to around $6,000. An experiment with nongovernment projects is well worth the money.

Congress has nurtured private transportation systems before, with airmail contracts for the airline industry and land grants for western railroads.

Now it's time to unleash the American entrepreneur in space to create a new orbital economy.

Congress has plenty of subsidy money available. It could divert some of the $25 billion it spends on agriculture subsidies to a purpose more suited to the nation's future. The nation has no shortage of farmers, but it desperately needs more space entrepreneurs. A modest, limited-term subsidy for space flights would be repaid a thousandfold and more.

Rockets and Spaceplanes of the 1990s   3

Today's expendable rockets cost less than the shuttle, but they're not cheap. The major aerospace corporations are charging more than $2,500 per pound for the Titans and Deltas, and the start-up companies are asking in the $2,000 to $10,000 range.

Driving prices lower is critical to encouraging the new space economy. Robust space development requires a fresh generation of spacecraft able to send people and cargo into space for $100 to $300 per pound.

These lower costs would permit the first passenger tickets to orbit, selling for about $100,000. As volume increases, they'd drop to $50,000 or even $30,000. These figures are far better than the $2 million that space shuttle tickets would have to cost, if they were available. (NASA vetoed space tourists in the shuttle cargo bay even before the *Challenger* disaster. However, the Soviet Union has sold a ticket to its *Mir* space station to a Japanese broadcast journalist for $11 million, for a voyage in 1991.)

Affordable spacecraft are taking shape now on computer-design screens in the United States, Japan, Britain, and West Germany. Some are innovative rockets and others are one-piece shuttles. With launch costs suitable for corporations instead of governments, they will finally end the test pilot stage of space travel.

A $50,000 ticket is too high for all but the wealthiest tourists, but it does fit corporate checkbooks. Companies will simply compare the ticket price to the potential profit. High-tech companies already buy expensive equipment for staff scientists, and many will consider buying some on-orbit time to help accelerate their research. In addition, universities with large endowments may start sending a few of their academics into space. Later chapters show how chemistry and biology work can be done more efficiently in space.

Academic groups are already active. One set has established the first International Space University (operating only during summer sessions at rotating host schools), and a university consortium is backing a low-cost space station (see Chapter 10).

Utilities may pay the $50,000 ticket price to start experimenting with prototype solar power plants to beam electricity down to their customers. A small hotel industry could start to serve the new arrivals, and from there it's only a short step to a tiny tourist trade. With the advent of the $50,000 ticket, space would finally become a productive extension of modern life.

Exactly *when* the new generation arrives will depend on public opinion. Congress will have to supply several billion dollars, with some going to NASA for research and the rest as incentives to private firms. The new vehicles could enter production as early as 1998 with vigorous help — or they could be delayed past 2020 without it. Their arrival depends on when government and corporate leaders seize the new opportunities.

Technology is not the stumbling block. Laboratory breakthroughs have far outpaced the money and political support needed to turn them into working hardware. With steady funding, NASA and the Pentagon could build a prototype spaceplane before the turn of the century, and U.S. industry then could build the commercial fleet. To stay within its overall budget, NASA would need to scale back its space station. NASA officials also would need to admit the shuttle is outmoded and should be retired. Otherwise, passenger tickets may be available on the advanced spacecraft of Great Britain, West Germany, or Japan but not the United States.

The shuttle's designers originally tried to reach space for $100 to $300 per pound, but had to make too many compromises and failed. The new space fleets could fail as well. But we've learned some lessons that will help the new effort:

— NASA can research new technology and prototypes, but private operators should design, build and run the commercial fleet.

— Everyone shouldn't be forced to use the same spacecraft. Specialized vehicles are needed; this will happen naturally when space transportation is provided by competing private firms.

NASA lives in a political world and succeeds using political strategy. NASA and the Defense Department both try to force Congress and the White House to accept their projects by making sure there are no alternatives available. That's why NASA shut down production of Saturn boosters (used in the Moon program and to launch the *Skylab* space station) long before it had a functioning shuttle.

NASA is taking the same approach on the space station. Only a $30 billion version was offered to Congress. (When first unveiled, the station's cost was put at only $8 billion.) A less risky approach would have started with a small module that NASA could use as a test-bed for later stations. NASA didn't offer this option, for fear Congress might decide the prototype was good enough and refuse to pay for the full-scale version.

NASA is now designing its $30 billion space station using simulations and projections, rather than the hard data that an interim project could have provided. The technical risk is higher, but the political risk of being stuck with the interim version forever is eliminated.

This attitude toward space projects guarantees expensive programs. NASA will never change because its environment will always be political. Rational choices can be made only when people are using private money. Government incentives can lift the overall feasibility of early space endeavors, but the *choices* must be made in the private sector. A savvy corporation, for example, would never try to create a $30 billion space station as its first step. It would find intermediate projects, with each step reducing the technical risk and broadening the market for space station services.

There might not ever be one, single space station. As Chapter 10 shows, NASA is trying to jam several incompatible uses into its space station. Corporations likely would create several space stations, specializing in the tasks each could do best.

Savings from the competition and specialization of private space development will be reinforced by the efficient new tools now available to space engineers. For example, wind tunnels can be replaced by supercomputers able to solve incredibly complex equations that predict air flow over advanced vehicles. Engineers won't need the expensive trial-and-error method of building scale models and laboriously testing each variation in a wind tunnel. The supercomputer will be the main tool, with wind tunnels only used to check the validity of new simulation programs.

At the other end of the computing scale, personal computers (PCs) give individual engineers unprecedented power. Automated drafting software on a PC makes one engineer equal to a small army of 1960s drawing assistants using pen and protractor to create mechanical drawings.

When the PC-created drawings reach the shop floor, they flow through another set of computers that control the manufacturing process. Shop workers consult PCs at each work station to see what steps are next in line. Scheduling is improved and labor costs drop. Because technicians are always working from the latest version of the blueprints—as shown on the PCs—quality control improves.

The rockets themselves will be built with super-strength metals and plastics that simply weren't available to earlier rocket engineers. The combination of new tools and new materials promises steep declines in the cost of space transportation.

SOMETHING OLD OR SOMETHING NEW?

Several designs are contending to be the next step in U.S. rocketry. The selection hinges on what goals are selected for U.S. space efforts. A small-scale space program can survive on rehashed shuttle and Titan rocket designs, but ambitious space station and Mars projects require something new. These are the three major paths available:

Updating the shuttle. The shuttle's capacity can be expanded through new boosters, a new external tank, and a cargo pod that would replace the orbiters on certain missions. NASA asserts these upgrades could be ready in the mid-1990s.

The Advanced Launch System (ALS). The ALS is a new expendable rocket for the late 1990s that would use very near-term technology in designs specifically chosen for lower cost rather than higher performance. It's intended for heavy lifting of space station parts, Mars expedition material, and the Strategic Defense Initiative.

The National Aerospace Plane (NASP). The NASP would be the very first craft to reach orbit in one piece and return intact. It requires very advanced engineering, has many skeptics in the bureaucracy, and would revolutionize space transportation and world politics. Prototypes could fly in the late 1990s if given substantial funding now. Later programs would use the NASP experience in creating transatmospheric fighter planes, affordable spacecraft, and supersonic commercial passenger jets.

People who think money will be scarce for space projects favor updating the shuttle. The ALS and the NASP get the backing of those who want to fight for a broad space frontier.

UPDATING THE SHUTTLE

NASA's shuttle can carry 48,000 pounds to orbit, far less than its original design goal. As a result, space station modules will have to be carried up empty and outfitted with equipment delivered by later flights. On-orbit assembly will be difficult and will monopolize astronaut time for many missions.

The shuttle could carry more payload if it used more powerful booster rockets. NASA wants to spend up to $1.5 billion creating advanced solid rocket motors (ASRMs) that would add 12,000 pounds to the shuttle's lift-off capability. The ASRMs would have fewer joints (the cause of the *Challenger* disaster), and the joints would be better designed. They'd also be built with a more automated plant to ensure higher overall quality.

NASA also could switch to liquid rocket boosters (LRBs) and get the same increase in launch power. LRBs can be test-fired before a launch, unlike solid rockets that burn all the way through once started. After launch, liquid rockets can be shut down or throttled up if needed to abort the mission. Solid rockets can't.

LRBs also would put less stress on the orbiters because they create less noise and vibration. The orbiters might last longer.

The shuttle could get additional lifting power from a lighter-weight external tank (ET). An ET made from an aluminum-lithium alloy would weigh 12,000 pounds less than the current all-aluminum tank. This translates almost directly into greater cargo capacity because the shuttle carries the ET almost all the way into orbit.

A final step would be creating a version that replaces the orbiter with an unpiloted cargo pod. NASA calls this "shuttle-C" for cargo carrier. Shuttle-C would boost 100,000 to 150,000 pounds into orbit, enough to carry a fully equipped space station module. The three-year, 19-flight schedule to get the space station into orbit with existing shuttle equipment would be cut in half.

NASA believes shuttle-C could be developed with only $1.5 billion and four years of effort. The agency sees the evolutionary nature of shuttle-C as an advantage, reducing development risks and speeding availability.

NASA asserts that shuttle-C launches would cost the same as piloted shuttle missions but carry three times the payload. This would bring the cost per pound down to $2,000, not counting the extra development costs.

The military, which has twice the space budget as NASA, sees the evolutionary nature of the shuttle-C as a drawback. When the next shuttle disaster strikes, they don't want the nation's heavy-lift booster grounded along with the piloted shuttle fleet.

The Office of Technology Assessment, a congressional group, also cautions that shuttle-C will be competing with regular shuttle launches for very overworked ground crews and launch facilities. Delays in the piloted shuttle schedule would shove back shuttle-C launches as well.

THE PENTAGON'S CHOICE

The Air Force favors a fresh start on heavy-lift boosters. Taking a clean sheet to rocket design may be able to force costs far below that of shuttle-C. The official goal is $300 per pound if production rates are high to support deployment of a strategic space defense. Without SDI cargoes, the system will struggle to force costs down to the $700 per pound level. Skeptics doubt the federal space bureaucracy will be able to reach that target.

A fresh start also would protect the heavy-lift launcher from being grounded by future shuttle disasters. Because another shuttle accident is almost inevitable, Air Force planners believe an entirely separate design for the heavy-lift booster is imperative.

NASA and the Pentagon are jointly studying this new booster through a series of ALS study contracts. The rocket may be completely expendable or include reusable engines like the shuttle. Its goal is to put 110,000 to 140,000 pounds into an equatorial low Earth orbit.

Three companies are working on designs to be completed in 1990. The rocket is to begin flying in the late 1990s. The ALS enjoys good support at the Pentagon, which sees it as much lower risk than the National Aerospace Plane. This lower risk makes it the likely choice as the only new U.S. spacecraft this century. The NASP is continually attacked as too aggressive, too uncertain, and thus too expensive to fund at a full-ahead rate.

The three ALS contractors are Boeing Aerospace, General Dynamics, and a team of Martin Marietta and McDonnell Douglas. Each of their designs would have a central core fueled with liquid oxygen and liquid hydrogen. Those fuels provide the best combination of zero pollution and lift at all altitudes.

The ALS will use boosters, just like the shuttle, that fire for about three minutes and fall away. General Dynamics proposes a booster identical to the core stage, so engines can be produced in high volume to bring costs down. The drawback is that oxygen-hydrogen engines aren't as powerful in the lower atmosphere, where the best fuel is a hydrocarbon such as methane, not hydrogen.

The General Dynamics booster would be recovered and used again just three or four times. By limiting the number of repeat launches, General Dynamics hopes it can spread each booster's cost over several launches without getting into expensive engineering and inspection problems of really long-term recycling.

Martin Marietta is considering an all-expendable rocket with liquid or solid boosters, or one with fly-back boosters. The fly-back boosters would cost less in the long run than boosters that have to be dragged out of the ocean. They have fairly high engineering risk, since none have ever been created.

Boeing Aerospace also has considered fly-back boosters, with aircraft-style engines strapped on to ferry the rocket back to its home base.

A key to the ALS is low-cost engines. The shuttle's main engines cost $30 million each and untold millions to refurbish after each launch. Now NASA is seeking expendable main engines costing only $3 million to $5 million. They will operate with combustion chamber pressures of 2,000 pounds per square inch (psi), down from the 3,200 psi of the shuttle main engines. The lower pressure allows engineers to specify much simpler pumps and other equipment surrounding the engine. Engineers also expect costs to drop by designing for fewer labor-intensive welds, and by using new materials that don't need expensive protective coatings.

The Office of Technology Assessment warns that NASA and the Air Force are concentrating too much on new hardware and not enough on launch operations. Finding ways to process and launch rockets with fewer people and less time is equally important to lower costs, OTA says.

THE NATIONAL AEROSPACE PLANE

The most controversial project in the U.S. space effort is the NASP. It would speed ten times faster than the supersonic Concorde, streaking along at the very edge of the atmosphere or zipping all the way to orbit.

Success rides on technologies never proved in practice, so it constantly draws attacks from hand-wringers who always doubt that any new science will actually work. Worse, the project's goal has been misunderstood ever since 1986, when President Ronald Reagan labeled it an "Orient Express" to carry passengers among Pacific Rim cities.

The NASP is an *experimental* plane and carries the X-30 designation to recognize this fact. It won't carry executives to Tokyo or do any other routine work. The Air Force and NASA are building the X-30 simply as a flying test-bed for advanced propulsion, cooling techniques, and materials that can't be tested on the ground.

The two X-30s now planned will be able to carry two pilots and a few thousand pounds of instruments. They will attempt to fly directly into orbit from a runway, without booster rockets or anything else that has to be replaced between flights. If successful, the X-30 will give engineers the confidence to design three different aircraft in a second generation:

A commercial spaceplane able to carry passengers to orbit at extremely low cost. A spaceplane will fly round-trips to orbit every day or two, so its construction cost will be spread over 100,000 passengers or more during a ten-year life. Tickets to orbit would be possible in the $10,000 to $100,000 range.

A space fighter-bomber-spy plane. Military versions would become a fleet of true war-fighting spacecraft, able to deliver bombs or spy cameras to any point in the world directly from U.S. airfields. Space fighter-bombers flying from Omaha wouldn't need vulnerable aircraft carriers or expensive foreign bases. They'd be able to take off from the United States and bomb Libya in under 45 minutes. Air Force generals believe that twenty-first century fighter pilots also could use spaceplanes to battle incoming nuclear warheads as part of a strategic missile defense. And they certainly would be the most capable spy planes ever created.

By basing space bombers securely in the United States, the military could abandon all its overseas air fields and all its expensive aircraft carriers. One space fighter armada in the United States could respond to crises anywhere in the world, lifting the burden of paying for separate forces in Western Europe, the Middle East, and Korea.

The Orient Express. Airline versions of the spaceplane would transport passengers from New York to Tokyo in a few hours. By flying very high, the Orient Express might even be able to cross over land without creating sonic booms.

The NASP would revolutionize both space travel and defense. At the same time, it would ensure that U.S. firms like Boeing continue to lead the world in passenger jets, which amount to a huge fraction of the entire American export effort.

With the X-30 so crucial to the country's future, it should get complete support from Congress, the Pentagon and the White House. Because it's only an experimental program, there's no danger of runaway expenses from trying to build an operational fleet.

The spaceplane's enormous military potential may allow the United

States to dispense with both aircraft carriers and foreign military bases. Aircraft carriers are sitting ducks for saturation missile attacks by any Third World dictator. Great Britain learned this in the Falklands War when Argentina used cheap Exocet missiles to sink an expensive destroyer. The United States suffered its own naval losses to cheap missiles in the Persian Gulf, but the lesson probably won't sink in until an aircraft carrier is destroyed while it "shows the flag" in hostile territory.

The military/congressional establishment is reluctant to admit that a space fighter-bomber would make aircraft carriers and foreign air bases obsolete. Admitting that the spaceplane is the answer would require admission that aircraft carriers and foreign air bases are problems. Mothballing the carriers and closing foreign air bases would save taxpayers $50 billion to $100 billion a year.

NASA is no better. Its strategists have decided the space station is to be the *only* major project of the early 1990s. They believe that Congress will support just a single big project at a time, if it supports space at all. Trying for two simultaneously seems futile to agency officials scarred by the budget wars over the shuttle. Only a more modest space station design would give NASA the cash to do both.

Government officials fear that demand for space transportation simply won't materialize in the late 1990s. They don't want to pour cash into full-scale spaceplane testing with the need for launch services wildly uncertain.

The overwhelming majority of launch customers have always been satellite communications firms. But that market is in turmoil because fiber optic cables are better and cheaper for many communications links. That's thrown launch forecasts into doubt.

Outside of communications, the need for launches is cloudy as well. In space manufacturing, firms and universities were just gearing up to test various processes and products when the *Challenger* disaster struck. The nearly three-year grounding of the fleet will be followed by two or three additional years where military and big-ticket NASA missions monopolize shuttle payloads. Almost all the flights dedicated to microgravity research have been canceled through 1990. Forecasters can't predict how long it will take for space manufacturing plans to bounce back.

In the military arena, a strategic missile defense may create a market for hundreds of launches during a five- or six-year deployment. But at any point, that market could disappear if there's a new arms control agreement or political support for missile defense collapses.

This all adds up to complete confusion over the demand for launches in the late 1990s. And without some forecast of the potential market, it's hard to predict what types of spacecraft will be best suited for it.

THE SPACEPLANE CHALLENGE

Space engineers have tried for decades to design spacecraft that reach orbit with a single stage (rocket style) or in a winged craft (spaceplane style like the NASP). Weight problems always got in the way. Engineers never could pack enough power into a single stage to reach orbit while carrying a substantial payload. The huge tanks needed for fuel and oxygen at lift-off become a real penalty near the end of the ascent—the last 5 or 10 percent of the fuel has to put the almost empty tanks into orbit along with the cargo. Rockets can lift vastly more weight into orbit by discarding empty tanks in stages as they rise.

Space engineers have naturally looked for ways to make smaller and lighter tanks. One method is to bring only fuel along, taking air from the atmosphere as oxidizer. Jetliners operate this way. The drawback to jet-style engines is their low power. They don't provide enough thrust to boost a spaceplane vertically like a rocket. A spaceplane will need wings to provide lift as the gradual jet-style acceleration builds up.

Even with wings for help, existing jet engines aren't powerful enough for a one-stage spaceplane. But new engines and lightweight materials make a true spaceplane possible soon, and it could slice costs to revolutionary new lows.

Jet-style engines for a spaceplane might be powerful enough to reach orbital velocity while still in the atmosphere. If not, a spaceplane could switch to rocket power for the final sprint into the vacuum of space.

A single-stage design brings tremendous savings in daily operations. As noted, standard rockets use several stages for efficiency—the bulky first stage fires only a few minutes and falls away to save weight. It's very hard to make such rockets totally reusable because they scatter their parts halfway around the world.

For example, NASA laboriously recovers the shuttle solid rockets that fall into the Atlantic Ocean. Considering the costs to refurbish them, this may or may not be cost-effective. No recycling is possible with the external tank because it is destroyed when it plunges into the Indian Ocean. A spaceplane won't drop key parts into the world's oceans—it reaches orbit and returns in one piece. This single-stage reusability is the source of the lower costs of spaceplanes: Nothing is thrown away, and nothing has to be fished out of the ocean and put back on.

That's the good news.

The bad news is that spaceplane research is emphasizing the military applications and not cheap civilian space transportation. Spaceplane research is 80 percent funded by the Pentagon. Some aerospace executives say it should be renamed the "Libyan Express" for its possible first military target. While the spaceplane will be a breakthrough military advantage for the United States, developing a space fighter-bomber is much harder than a civilian space transport. By skewing X-30 research to support primarily military goals, civilian use of spaceplane technology will be needlessly delayed.

Many basic facts about the U.S. spaceplane aren't available because the Pentagon classified all the contracts and told the five winners to keep quiet. The secret contracts call for the design, construction, and ground test of the crucial new engines, the airframe, and other components. In May 1990, the Pentagon will order construction started on two experimental X-30 spaceplanes, aiming at the first flight in 1995. Total military and NASA spending before that first flight is expected to be $3 billion.

Commercial spaceplanes and a ground-to-ground Orient Express are expected to lag many years behind the military fleet. There's no technical reason for that; in fact, a civilian spaceplane would be much easier to engineer than a warplane. A commercial spaceplane simply goes up and returns, while a warplane must be capable of highly stressful maneuvering and have much greater long-distance endurance. The military will get its spaceplanes first because the Pentagon has been the only organization willing to spend the development money. Defense Secretary Richard Cheney tried to kill Pentagon spending for the NASP in 1989, however. Cheney proposed tossing the project to NASA, where it would shrink to a purely long-range R&D effort with no plan to build actual prototypes. NASP has powerful supporters and likely will retain at least some level of Pentagon funding.

White House officials have strongly supported the military spaceplane. George Keyworth, when he was science adviser to President Reagan, was the project's most visible backer. Officials in the National Security Council (NSC) continued cheerleading for the spaceplane after Keyworth left the administration. Gerald May, the NSC's director of space programs, told an industry conference in late 1986 that the spaceplane "really looks do-able."

Vice President Dan Quayle, the head of the National Space Council, also has praised the project as vital to national interests.

Robert Williams, who ran the spaceplane office at DARPA until recently, is equally convinced. He was challenged at that conference about his optimism, given NASA's caution on spaceplanes. NASA's director of advanced programs says single-stage spaceplanes are decades away because they require an astounding 60 percent increase in engine and component performance.

"We've *already* measured greater than 60 percent improvement in ISP (engine thrust) and materials," Williams responded. An air-breathing spaceplane can put 1 pound of cargo into orbit using only 0.25 pounds of fuel, according to Williams, while a rocket engine needs 2.4 pounds of fuel to accomplish the same thing. Pentagon experts see no reason to dawdle on spaceplane work, and they've plunged into development at full speed.

The program has ten expensive NASA and military supercomputers working almost full-time on fluid dynamics. The supercomputers use complex equations to predict how air will flow around the wings and through the engine. Using ten supercomputers simultaneously is a massive assault on the key design questions of a spaceplane, amounting to 30 to 35 percent of the government's entire stock of supercomputers.

The X-30's skin will be made from exotic new metals that are much stronger than today's aluminum coverings. One potential new skin material is powdered titanium metal, impregnated with ceramics. The powder is placed in molds to create the right shape. The titanium and ceramic particles are then heated but kept below titanium's melting point. They bond to each other in their original "fluffy" state, full of holes. Parts made with this type of powder metallurgy weigh much less than solid parts but resist heat even better because of the ceramic particles embedded in them.

A key recent advance is the creation of extremely fine powder through rapid-solidification-rate (RSR) casting. A spray of melted titanium is flash cooled, before the tiny droplets have time to clump together. The superfine RSR titanium powder results in greater strength for the eventual product.

Another candidate is titanium or aluminum melded with a mesh of silicon carbide fibers. Strands of silicon carbide are extremely light—a spool of 30,000 feet weighs only one pound. When added to low-cost aluminum, these fibers quadruple the metal's strength. Reinforced aluminum becomes stronger than titanium, a vastly more expensive metal now used in the most advanced applications.

When the fibers are added to titanium, they create a metal that can maintain its strength and stiffness even when heated beyond the 1500-degree Fahrenheit level. That's the temperature expected on the NASP's skin during its streak to orbit. Fiber-reinforced aluminum may be used for internal NASP parts, reducing costs, with reinforced titanium reserved for the skin.

Progress made by the supercomputers and by new metals make the planners confident the X-30 will fly by 1995.

Improved metals are crucial to the military spaceplane. It may build up tremendous heat by plunging quickly into the thick lower atmosphere to bomb or photograph a target. Commercial spaceplanes can return on a gentle reentry path, one with a long glide in the upper atmosphere. They can slow down gradually and take time to radiate the excess heat into the atmosphere.

Dipping into the atmosphere to drop bombs or take pictures and zooming back into space also puts big demands on the engines and adds to the fuel load required by a military plane. Swooping down and returning to orbit is not something a civilian craft will need to do. The military has other unique requirements as well:

— The ability to take off under all weather conditions on short notice.

— The ability to evade radar detection and heat-seeking missiles by having "low observables." This stealth capability will require special shapes and materials that diffuse and absorb radar beams instead of reflecting them back to the enemy.

— Strength to resist damage from laser beams or particle beams.

These features will create a powerful military fighter-bomber or almost invulnerable spy plane. But many of the military needs are excess baggage to a commercial spaceplane.

The biggest challenge in the X-30 program is faced by the two firms — Rockwell's Rocketdyne division and Pratt & Whitney — that must create the new engines.

Today's airline and fighter engines are turbojets. These use a fan-like compressor to pump air into a combustion chamber, where it mixes with fuel and burns to provide thrust. Turbojets power the Concorde to Mach 2 — two times the speed of sound — and the Air Force SR-71 reconnaissance jet to Mach 3.

The X-30 needs far faster engines. The engines proposed for the spaceplane eliminate the mechanical compressors, because they simply can't survive at higher speeds. The spaceplane engines will rely on the supersonic forward motion of the craft to ram air into the engine. This principle has been used in ramjets, where the incoming air is forced to slow to subsonic speeds in the combustion chamber to give the fuel time to burn.

Figure 3.1
The X-30 Spaceplane

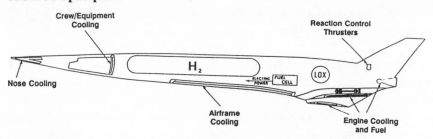

Source: NASA.

Ramjets can power a plane to Mach 5 or 6 but not beyond. The limit is
set by heat. In a ramjet, the incoming air slows down as it enters the combustion chamber. The kinetic energy of its speed is converted into heat
energy as it slows. Beyond Mach 5 or 6, the conversion of kinetic energy to
heat energy creates temperatures so high they melt the engine.

Recently, scientists discovered how to burn hydrogen at supersonic
speed. The incoming air doesn't slow down on its way through the engine,
which removes the Mach 5 speed limit and creates the supersonic combustion ramjet, or scramjet.

The hydrogen also will help keep the X-30 from burning up as it plunges
through the atmosphere at breakneck speeds. Liquid hydrogen is stored at
-423 degrees Fahrenheit in the X-30's main tank. Before the supercold
hydrogen goes to the engines, it flows under the X-30's skin to drain off
heat. It also cools the crew cabin and key engine parts before it squirts into
the combustion chamber.

The unclassified sketch shown in Figure 3.1 is a general view of the X-30
and illustrates how it will use hydrogen for cooling. The X-30's *actual* appearance may be quite different. Artist's concepts available early in the program were deemed too specific and are no longer released.

Making the fuel do double duty as refrigerant is typical of how other
X-30 parts have multiple roles. The front part of the body, for example, will
help compress air as it flows to the engines. The X-30 body may resemble
an ice cream cone flying point first. (That's the way it looked in an early
artist's concept.) The engines will be mounted in an arc about two thirds of
the way back, about where the cone would end and the ice cream would
start. The increasing width of the cone would compress air as it approaches
the engines. After combustion, the rounded rump of the plane would serve

Figure 3.2
Engine/Body Integration on the X-30

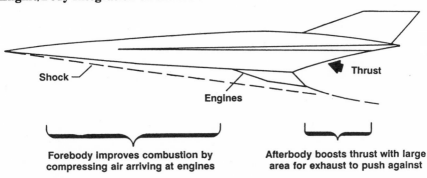

Source: NASA.

as an extension of the engine. Air expanding from combustion would push against the rump to give the plane an extra shove.

The drawing in Figure 3.2, an unclassified sketch released by DARPA, shows one way the compression would work. Because the engine design is top secret, the sketch probably disguises the real shape of the X-30.

One of the mysteries of the X-30 is how it will get up to supersonic speed where ramjets and scramjets work. The plane possibly could carry auxiliary rockets for the initial kick, or the rocket capability might be built into the ramjet/scramjet engines. Leaders of the project aren't saying what the solution will be.

Defense officials were much more talkative about scramjets when they were still looking for money. Back in December 1984, former DARPA chief Robert Cooper revealed that scramjets might not need any rocket assist on takeoff. Cooper said a DARPA study showed that an aerospace plane could take off from a 10,000-foot runway using scramjets, inject itself into orbit, dip into the atmosphere more than once to take spy photos or drop bombs, boost back into space, and come home to land on a standard runway.

DARPA officials also indicated to Congress in 1985 that scramjets might not be the only way to go. DARPA said it was developing "airturbo-ram-jets" that could handle thrust from zero to over Mach 5, which would break the Mach 3 barrier imposed on today's turbojets by turbine problems. The turbine is the weak link because it's under the most stress. It's located inside the combustion chamber where hot exhaust gases spin it at high speeds to provide power for the compressor. Above Mach 3, the double stress of combustion heat and spin destroys the turbine blades.

In the proposed airturbo-ramjets, the turbine is removed from the combustion chamber. The turbine spins from the push of a separate gas generator, running at much cooler temperatures than the engine. The switch takes a major chunk of sensitive equipment out of the plane's harshest environment. The streamlined combustion chamber is freed to run at much higher speeds and temperatures. The stationary walls of the chamber can stand the higher heat because they aren't under the spin stress that the turbine blades were.

DARPA says airturbo-ramjets can boost a spaceplane from the runway up to Mach 5 without trouble. That's all a commercial Orient Express from New York to Beijing would need. If Boeing, McDonnell Douglas, and Airbus Industries had the cash and inclination to work on airturbo-ramjets, they could build jetliners more than twice as fast as the Concorde without waiting for the military to perfect the X-30.

A slightly different model of those jetliners could carry a civilian spaceplane piggyback to great speed and altitude. There it would separate and fire rockets to make the final push for space. The investment in the jetliner's development would serve two different markets: (1) high-speed passenger service on Earth, and (2) cheap lifts to orbit.

Germany's proposed Sanger spaceplane (covered later in this chapter) takes this piggyback approach. The Pentagon's current opinion is hard to know, because recent testimony to Congress has dropped all mention of airturbo-ramjets, along with any details on scramjet performance.

The Pentagon likely will skip airturbo-ramjets and concentrate on the more exotic scramjets because they can push a plane all the way to Mach 25. Scramjets also will give military spaceplanes tremendously long ranges, perhaps being able to circle the globe several times while changing directions and altitudes repeatedly. An airturbo-ramjet that needs rocket power to reach the highest speeds might be able to reach orbit or go city to city, but wouldn't have the range needed for military missions.

Concentration on military missions means less challenging civilian options aren't being pushed. NASA's spaceplane manager Lana Couch, for example, believes that air-breathing engines capable of hitting Mach 12 are realistic based on what NASA's already learned. (Mach 12 is twelve times the speed of sound.) Doubling that to Mach 20 or 25 to reach orbital velocity is what requires the special X-30 research spaceplanes. Commercial spaceplanes might be able to fly quite economically up to Mach 12 and then switch to rocket power, bypassing the lengthy process of building the X-30 research planes.

As an interim vehicle, a civilian spaceplane might not use air-breathing engines at all. NASA and European researchers say that substantial improvements can be made in rocket vehicles, without the go-for-broke risk of creating an all-new air-breathing plane. Two experts from NASA's Langley Research Center, Charles Eldred and Theodore Talay, gave this analysis to an international conference in 1986:

"The analogy of airplane-like operations is often used as the goal for advanced fully reusable launch vehicles. . . . The airplane analogy should, however, be used with discretion because of the fundamental differences between the classical airplane and the launch vehicle.... The transport airplane is a cruise vehicle designed for range, whereas the launch vehicle is an accelerator with orbital speed as the objective. Concepts which work well for cruise may not work well for accelerators and vice versa."

Eldred and Talay note that vertical takeoff gets a vehicle quickly out of the thick atmosphere. The quick exit means there's not much problem with heat from atmospheric friction on takeoff. Airplane operations—with the vehicle horizontal—do offer easier servicing and cargo loading, but the two experts have an answer to that as well: An empty liquid rocket can undergo ground processing while conveniently horizontal, and then be turned upright and fueled.

The Langley experts proposed a fully reusable rocket plane with detachable engines and payload carrier (Figure 3.3). The removable engines allow easy servicing, and the tanks for supercold liquid hydrogen and oxygen can be thoroughly inspected for cracks. The detachable payload bay is made possible by the "flatbed truck" shape of the vehicle's top surface. It has no central vertical fin, unlike the shuttle orbiter and most commercial aircraft.

Figure 3.3
Rocket Spaceplane

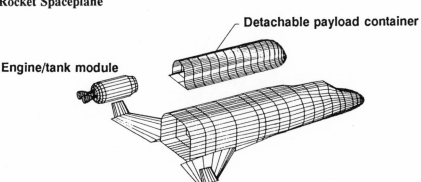

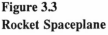

Source: NASA.

Figure 3.4
Comparison of New Rockets to Shuttle

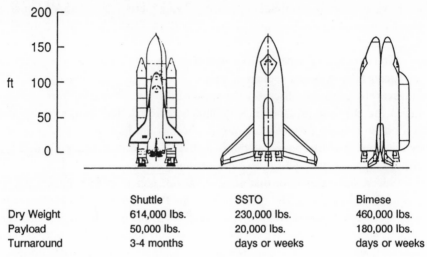

	Shuttle	SSTO	Bimese
Dry Weight	614,000 lbs.	230,000 lbs.	460,000 lbs.
Payload	50,000 lbs.	20,000 lbs.	180,000 lbs.
Turnaround	3-4 months	days or weeks	days or weeks

Source: NASA.

Small fins on each wing tip take its place.

The removable payload bay allows quick turnaround times on the ground. Fresh cargoes are simply snapped into place. Fast turnaround is the *key* goal of a reusable spaceplane – it's the only way to get enough flights per year to really cut costs.

Figure 3.4 compares the shuttle with new rocket spaceplanes. The SSTO stands for a "single-stage-to-orbit" system and Bimese refers to two spaceplanes that are joined, with one acting as a booster during the initial minutes of flight. The much shorter turnaround times of all-liquid rocket spaceplanes would revolutionize launch costs.

The dry weights refer to the weight of the vehicles before fueling. Engineers use dry weights as rough estimates of development and construction costs – the greater the dry weight, the more costly the development. The 230,000 pounds dry weight estimated for the SSTO rocket shows it might be roughly one-third the cost of the shuttle to develop and build. The Bimese option, since it would mate two almost identical units, should still cost much less than the shuttle to develop. But it would be able to launch almost four times the weight every few days, instead of every few months like the shuttle.

A SPACEPLANE WITHOUT WINGS

The vertical-takeoff rocket and the horizontal-takeoff spaceplane each have their advantages. The rocket gets out of the atmosphere quickly, without building up heat and wasting energy fighting air friction. The spaceplane is lighter weight because it uses the atmosphere for oxidizer. The best of both could be combined into one vehicle — launching straight up but using air-breathing scramjets for its main thrust.

Figure 3.5
Vertical-Takeoff Spaceliner

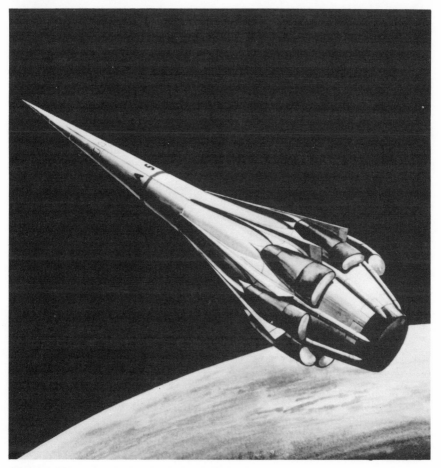

Source: L. J. Skubic (artist), The Marquardt Company.

William Escher of the Rocketdyne Division of Rockwell International proposes a design for this crossbreed spacecraft. It resembles the needle-nosed spaceships seen in 1950s science fiction movies (see Figure 3.5).

Escher's spaceliner would use combined-cycle engines. The scramjets would be combined with the rockets needed for initial lift-off. Air-breathing jet engines are too weak to blast off vertically all by themselves. That's why most scramjet concepts include wings, to give the engines time to gradually build up the vehicle's speed to escape velocity. Escher eliminates the weight of the wings by using rockets at takeoff.

After pushing the spacecraft to Mach 2 or 3, the rockets shut down, and the air-breathing ramjet continues. At Mach 6 the spacecraft shifts to scram-jet mode. It keeps accelerating and rising until the air becomes too thin for the scramjet. The precise point where the scramjet quits depends on how well it has been designed, but it will be somewhere between Mach 12 and Mach 25. Then the spaceplane switches to rocket propulsion for the final push into orbit.

The return trip is perhaps the hardest part. The ship flies back nose first on ramjet thrust. Just before landing, it must flip over to land upright with a final burst of braking power. The abrupt change to bottoms-first flight is risky when the spacecraft is close to the ground and almost out of fuel.

Escher's combined-cycle spaceliner places the main payload compart-ment at the very bottom. Cargo can be rolled right into the spaceliner at ground level. The engines are located in a band around the spaceliner's "waist." The ground-level cargo bay and the high-mounted engines save a lot of money on ground facilities. No launch tower is required when crew can simply walk in at street level, and engines placed at the ship's middle eliminate the need for expensive flame-diversion ducts in the launchpad. Launch site costs aren't a trivial matter—the Air Force spent $3.3 billion building a West Coast shuttle launch center. Private operators can't afford that level of investment, so the simple requirements of Escher's spaceliner have great appeal. A vertical takeoff and landing spaceliner has other ad-vantages:

— Eliminating the wings reduces weight and aerodynamic drag.

— Vertical launch and landing eliminates the need for a retractable landing gear system, saving more weight.

— Scramjets need help from the entire shape of the spacecraft to work ef-fectively. The forward half of the craft must help compress air before it goes into the engines, and the back section must be an extension of the nozzle. A winged spaceplane with engines only on the bottom loses any help from the top surface of the craft. A vertical-takeoff spaceliner can use a ring of engines that completely use the whole craft's shape.

— The spaceliner's circular shape is the most efficient for fuel tanks. Circular tanks hold the most fuel for the least amount of metal walls. The squashed shape of winged spaceplanes requires more material for tank walls to hold the same volume of fuel.

— The circular shape will cost less to build. Circular walls and braces that repeat throughout the spaceliner will be much easier to mass-produce than the complicated curves of a winged ship.

— Escher's spaceliner can make emergency landings on any firm, level site. Winged spaceplanes must try to reach major airports or military bases with very long runways.

The vertical-launch spaceliner needn't stop after reaching low Earth orbit. Refueled, it could carry people and supplies to the high orbit of geosynchronous communications satellites, or it could even continue on to the Moon. While it wouldn't be the "perfect" engineering design for Moon service, it could be the best business solution. Operators could skip the entire development process for a specialized space-only vehicle. They'd also gain from the mass-production efficiency of building one basic spacecraft model to serve both markets. The combined savings could amount to hundreds of millions of dollars, compared to the cost of designing and building separate winged and space-only vehicles.

The choice between winged and vertical-takeoff spaceliners won't be made for several more years. Winged spaceplanes have the public spotlight and virtually all the development money. But the true "best choice" may not be the one getting the most attention.

Right now, any commercial version of the spaceplane is remote because civilians are getting so few details on spaceplane research. Many experts bet the spaceplane will slowly fade into a completely "black" military program — black programs are so heavily classified that only a few members of Congress are allowed to see their budgets.

If Americans demand a separate civilian spaceplane/spaceliner program, we might attain cheap access to space before the twenty-first century. But business as usual offers no hope for commercial spaceplanes before 2020.

Figure 3.6
The British HOTOL Spaceplane

Source: Photograph courtesy of British Aerospace (Space Systems) Ltd.

THE BRITISH OPTION

The only active efforts to create commercial spaceplanes are in the United Kingdom and West Germany. British Aerospace is trying to rouse European enthusiasm for its single-stage HOTOL—Horizontal Take Off and Landing—craft. And in West Germany, MBB-Erno is pushing its two-stage Sanger proposal.

The British HOTOL's mission is simple: boost one commercial satellite at a time into low Earth orbit with an unmanned aerospace plane. British Aerospace believes the HOTOL can put payloads into low Earth orbit for just under $300 per pound, compared to $6,000 per pound for today's true shuttle costs. HOTOL would be slightly longer than the Concorde and would take off with about 50 percent more acceleration than a Boeing 747. A small fin on the nose makes HOTOL look something like a harpoon (see Figure 3.6).

HOTOL would take off on automatic pilot from a standard runway. Artificial intelligence and robotics would keep the plane flying whenever it was out of range of ground stations. Air-breathing engines would push it to Mach 5 and 15 miles high; then rockets would send it up 50 miles to orbital speed. It coasts from that point into a low orbit 180 miles above the Earth.

Although a crew could ride HOTOL's later versions, the initial vehicle would be a cargo craft. Eliminating the crew lets HOTOL skip the intensive safety tests needed for manned craft. This will get HOTOL into operation years earlier than a crewed spacecraft. Experience gained flying cargo will work out the bugs in the system without risking human life. Later efforts to qualify the HOTOL for crewed flight would be made with confidence gained through years of practical cargo use.

British Aerospace plans a modest goal for HOTOL's payload capacity. It would carry only 7 to 11 tons, about one third of the U.S. space shuttle's capacity. Each satellite customer would get a dedicated flight. Dedicated flights are far more attractive to customers than the shared flights of the U.S. shuttle, where launch dates often shift about as planners seek to fill up the payload bay with the best mix of cargoes.

The Mach 5 goal for HOTOL's air-breathing engines is much lower than the Mach 20 to 25 sought in the U.S. military aerospace plane. "I lose my nerve above Mach 5," explains Peter Conchie, who leads the HOTOL program for British Aerospace. The Mach 5 goal doesn't require the development of scramjets. It's within the range of the airturbo-ramjet, which uses relatively familiar techniques.

Rolls-Royce would build the HOTOL engines. The company is a world leader in turbojet engines for airliners and military jets. Secrecy cloaks the exact nature of the engines; the British feel the engines are a significant breakthrough, and they don't want competitors to get a head start. The "secret engines" were partly revealed in early 1988, when Rolls-Royce released a simplified diagram.

The HOTOL engine is a combination of rocket and air-breathing engine. It apparently cools incoming air, compresses it, and feeds it into a combustion chamber where it burns with incoming hydrogen. This works up to Mach 5, and then the rocket switches to tanked oxygen for the final burst to orbit.

The HOTOL engine cools the incoming air using the frigid liquid hydrogen that later will burn for thrust. The liquid hydrogen also is circulated around the combustion nozzle to keep it from melting. In the process, the hydrogen picks up heat energy that boosts its efficiency in the combustion chamber.

Because the HOTOL engine is air breathing only to Mach 5, it avoids the complex problems of supersonic combustion. The U.S. scramjet engines, by contrast, are requiring six supercomputers to design because they are so exotic. The U.S. plane also may need expensive new wind tunnels capable of testing spaceplane engines at speeds greater than Mach 7, the limit of existing wind tunnels. Thus, the U.S. spaceplane may achieve greater efficiency at the cost of a vastly more expensive and time-consuming development program.

HOTOL would be simpler than the U.S. aerospace plane and the U.S. shuttle in several ways:

— The payload bay would be nestled in the center of the plane, with a large hydrogen tank forward and a smaller oxygen tank and engines behind it. This placement would produce a good center of gravity that can't shift much, no matter where a cargo is loaded in the payload bay. The U.S. shuttle is the opposite. It becomes unbalanced unless NASA makes very careful calculations on where cargoes are stored in the payload bay. The sensitivity stems from throwing away so many components during the climb to orbit. The payload bay suddenly becomes a very large fraction of the shuttle's overall remaining weight once the external tank and solid rocket boosters drop away.

  NASA makes elaborate computer projections for each flight on how to fill the payload bay and produces customized flight software for the combination of payloads to be flown. The HOTOL will stay massive throughout the flight and won't need an army of programers and analysts to model each launch.

— HOTOL is designed to create less heat on reentry than either the shuttle or military spaceplane. The shuttle drops almost like a brick when returning to Earth because its stubby wings provide very little lift. HOTOL's wings will be much larger. On a pound-for-pound basis, HOTOL will have 2.5 times the surface area of the shuttle. It can reenter on a gentle slope high in the atmosphere where the thinner air slows the vehicle more gradually, building up lower heat loads. Peak temperatures on the HOTOL's skin will be several hundred degrees lower than on the shuttle, which allows HOTOL to use existing materials for its skin: a high-temperature nickel alloy on the belly and titanium on the top. HOTOL thus avoids the heavy and fragile ceramic tiles of the shuttle.

The American X-30, by contrast, depends on exotic new materials never before used in aircraft and on complicated schemes to refrigerate the skin by pumping liquid hydrogen in pipes under the outer shell.

British Aerospace says the first test flights could begin in 1996, with operational flights by the year 2000. The company wants to fly 30 times a year with a fleet of four HOTOLs. The ground turnaround time is very ambitious: only two days.

Manned flights could start early in the twenty-first century, when HOTOL also could serve as a civilian airliner carrying 60 people in the payload bay. London to Sydney flights in 67 minutes would be possible. HOTOL passenger flights would remain controlled by automatic pilots using artificial intelligence. The on-board crew could punch in coordinates for new orbital positions or landing sites, but they wouldn't actually be flying the plane.

British Aerospace is confident of HOTOL. Conchie told an industry conference in late 1986: "I'm still slightly amazed at how easy it is."

While Conchie was confident, his government was not. In 1988, the Thatcher regime refused to move HOTOL toward development. In response, HOTOL inventor Alan Bond formed a consortium of individual and institutional investors called the HOTOL Development Corporation. They promised to fund $250 million of research and development.

Roger Lanstaff of British Aerospace told a University of North Dakota conference in 1988 that the investors were willing to spend an additional $10.9 billion. Such an investment would move HOTOL into full-scale development, assuming the research shows sufficient results.

WEST GERMANY PUSHES A TWO-STAGE SPACEPLANE

The West German aerospace establishment doubts the HOTOL claims. Their research shows that a single-stage spacecraft is very vulnerable to development problems. If the engines aren't as powerful as expected, or the machine turns out to be heavier than first hoped, it won't be able to deliver a worthwhile cargo weight to orbit.

The German solution is a two-stage spaceplane. In 1986, researchers

Figure 3.7
The Sänger Spaceplane

Source: Photograph courtesy of MBB.

told the International Astronautical Federation that two-stage designs can better withstand development problems. Shortcomings in the first stage drop away when the second stage ignites. If the first stage proved too heavy, for example, that doesn't plague the craft all the way to orbit.

German research also showed that for single-stage spacecraft, a vertical rocket-style launch is most efficient. And that's poison in Europe. German aerospace executives say Europeans never will allow a rocket-style spaceport in their backyards. Only spaceplanes that take off horizontally like standard jetliners will be politically acceptable at European spaceports. Local spaceports are essential to low-cost operations; Europeans don't like shipping satellites halfway around the world to South America for Ariane launches in French Guiana.

MBB-Erno, the leading German aerospace contractor, calls the first stage of its proposed spaceplane the Sänger after one of its early engineers. The Sänger is larger than the HOTOL, measuring about the same length as a Boeing 747 and weighing considerably more. Sitting on top of the Sänger would be an orbital stage, called the the HORUS for crewed version and the CARGUS for the cargo-only version (see Figure 3.7).

The Sänger may be cheaper to develop than the HOTOL. MBB-Erno engineers argue that the Sänger first stage will build on the technology developed for the Concorde. And because the first stage can do double duty as a Mach 6 "Orient Express" for the airline market, its $6.6 billion development cost can be spread over two applications.

The second stage will build on the technology of Hermes, a manned capsule now under development for the Ariane rocket and described in the following section. That may drop its development cost to as little as $1.6 billion. The second-stage HORUS would carry two crew and almost 9,000 pounds to orbit, or it could carry ten passengers and 4,400 pounds of cargo. The all-cargo version might carry 13,000 to 26,000 pounds.

In 1989, the West German goverment granted MBB $200 million for a three-year program to prove the plane's feasibility.

The British and German engineers deride each others' plans. The British say the Sänger will bankrupt Europe with high operating costs. The Germans scoff at the "secret engines" of the HOTOL and doubt they will be as powerful as advertised.

EUROPE READIES THE HERMES SPACEPLANE

Europe is working on a $4.5 billion spaceplane project that may reach orbit shortly after the American X-30. Europe's three-person Hermes shuttle will ride atop an expanded version of the Ariane rocket. The combination of Hermes and Ariane won't be especially cheap to operate, because they are simply extensions of current technology.

The European Space Agency (ESA) says Hermes won't launch satellites as the U.S. shuttle does. Instead, the Hermes shuttle will be a taxi to deliver passengers to U.S. and Soviet space stations, and to a European free-flying platform—a microgravity laboratory that is to be visited periodically for crewed experiments. The twice-a-year Hermes flights also will service satellites. Crews will repair or retrieve damaged satellites and will refuel operating satellites. (Comsats need fuel for the small maneuvering rockets that keep them in the proper orbit.) Refueling lets designers devote more weight to useful electronics and less to station-keeping propellant tanks.

The Hermes shuttle will be the first new Western manned spacecraft in almost 15 years. (American firms like Martin Marietta and AMROC are aiming strictly at the cargo market.) While Hermes won't use any spectacular new technology, it still may be modestly cheaper than the U.S. shuttle because it has a narrower role: space taxi and repair truck. It won't launch satellites, and it will be too small to deliver prefabricated sections for a space station. Specialization will let the Hermes be compact and less complex than the U.S. shuttle, and that may reduce costs.

European planners hope an uncrewed Hermes will be flying on Ariane rockets by 1998, with piloted flights by early 1999. In addition to European countries, Canada is chipping in with 0.5% of Hermes' cost.

JAPAN PROBES SPACEPLANE TECHNOLOGY

Japan is the final player in the spaceplane race. Japan has a tiny rocket industry, where it is slowly gaining aerospace skills. But it lacks experience building large jetliners, and officials want Japan to enter that market. Japanese officials also see space as a key market for the next century, and they expect spaceplane research will help Japan be competitive in both jetliners and space.

Until recently, Japan launched copies of the U.S. Delta rocket, licensed from McDonnell Douglas. Now Japan is gradually replacing each stage of the Delta copy with its own domestic sections. The first step toward Japanese space independence is the H-1 rocket. It uses the Delta pattern only for the first stage. The American upper stage has been replaced with a more powerful design created by Mitsubishi Heavy Industries and IHI Inc. McDonnell Douglas likes the new upper stage so much it wants to put the Japanese technology into the American Delta to get more lifting power.

In 1992, Japan will go a step further and replace the Delta first stage, creating an entirely domestic H-2 rocket. It will be slightly more powerful than the Titan III rocket but will weigh only a third as much.

In 1996, the National Space Development Agency of Japan will followup with an unmanned orbiter called HOPE that will ride the H-2 into space. HOPE will give Japan a way to retrieve satellites and space experiments, including materials processing done on NASA's space station. The orbiter's main goal, however, is to give Japan the flight experience needed to create a crewed spaceplane early in the twenty-first century.

The Japanese Space Activities Commission recommends that Japan spend almost $14 billion on a crewed spaceplane. The panel also thinks Japan should build its own space station by 2000.

To get expertise in jetliner design and construction, which will lead to its spaceplane, Japan is negotiating with Boeing to jointly develop a new generation of aircraft. Japan would contribute its electronics and its knowledge of composite materials. On the spaceplane front, Japan's Science and Technology Agency has started distributing research contracts. Mitsubishi and Kawasaki Heavy Industries will study airframe design, while Nissan Motors and others will research scramjet and turboramjet engines. Fuji Heavy Industries will create an unmanned research vehicle, and Fujitsu Corp. will work on advanced control systems.

THE OUTLOOK

Current U.S. space policies won't produce a civilian spaceplane until well into the twenty-first century. A heavy-lift rocket — the Advanced Launch System — may be developed a little earlier, in the late 1990s. That will cut launch costs modestly.

Massive savings require some type of fresh start. That could come if a private firms like AMROC or Pacific American Launch Systems succeed with their maverick designs, or if popular demand forces Congress to invest significant money in *civilian* spaceplanes.

The $5,000 Ticket to Space   4

Customers of first-generation spaceplanes will pay with corporate checks. The second generation of spacecraft, however, finally will deliver space to the rest of us. Economy-class tickets are quite likely sometime early in the twenty-first century.

Popular-price transportation may begin with laser beam propulsion. Laser-boosted ships may drop costs to under $20 a pound. That's about $5,000 for a passenger seat to the new frontier. Anyone with the desire will be able to emigrate to the orbital factories and research stations of 2020 or 2030.

Laser propulsion overcomes the biggest drawback of rockets: the need to carry a fuel supply with them most of the way to orbit. Rockets are like a car built to go coast to coast on a single tank of gas. The car's gas tank would be the size of a Winnebago, and most of the car's initial efforts would be wasted just trying to move 30 tons of fuel. Rockets face the same problem as the coast-to-coast car.

Laser-boosted rockets get their efficiency by refueling as they make their ascent. Most of their energy will be beamed to them by laser as they rise.

Tapping into regional power grids at night, when the overall demand is low, will provide the massive surge needed for the laser-boosted rockets. The incoming beam will vaporize a small amount of fuel carried onboard to create superhot gas for thrust.

The gas is far hotter than ordinary rocket exhaust, so it provides much more thrust per pound, which means much less fuel — and weight — has to be carried into orbit. The fuel could be water, which would be blasted into individual oxygen and hydrogen atoms by the incoming laser beam. Some proposed laser ships would use graphite for the target fuel; some would use liquid hydrogen. Whatever the fuel, the real energy comes from the ground-based lasers, fed by cheap electricity.

Engineers began speculating about laser ships in the 1960s. The barrier was poor laser efficiency. Even in the late 1970s lasers could convert no more than 1 or 2 percent of the incoming electricity into laser beams. Now the tremendous research blitz under way for the Strategic Defense Initiative is creating lasers that are ten times more efficient.

The SDI work most useful for space travel centers on "free electron lasers," or FELs. Normal lasers work by pumping energy into a crystal or gas, forcing the electrons in atoms to higher orbits. When the electrons fall back down to their regular orbits, they spit out photons, or light waves. The problem is that raising the electrons into higher orbits requires much more energy than is released.

The FEL uses electrons that have been stripped away from atoms. A powerful stream of these "free" electrons is shot through a magnetic field, which "wiggles" them. This causes the electrons to convert some of their energy into photons, creating a laser light beam. The flowing stream of electrons then recirculates, returning most of its energy to continue the cycle. In addition, FELs can be tuned to produce any frequency of light from deep infrared up to ultraviolet and beyond.

The ability to tune the laser beam to particular frequencies (colors, if the frequencies are in the visible light range) is very important. For laser propulsion, the beam must travel hundreds of miles through the atmosphere, and air is more transparent at some frequencies than others. These transparent frequencies are called windows. Shooting through a window allows laser light to penetrate longer distances with less energy loss. SDI scientists say the best windows are at low frequencies, in the infrared range.

FELs are expected to be about 20 percent efficient in converting incoming power to laser beam energy. That's a revolutionary advance over normal gas and crystal lasers that struggle to achieve 1 percent efficiency.

Twenty percent is the "wall plug" efficiency, the most conservative way to rate lasers. One laser scientist says wall plug efficiency is calculated by dividing the heat in your laser beam by your electric bill.

Even higher numbers are tossed out by SDI enthusiasts. They can be misleading because they are based on different notions of "efficiency." In 1986, Lt. Gen. James Abrahamson, the first head of the SDI program, said that Lawrence Livermore Laboratory had reached 42 percent. These numbers are extraction efficiencies. They measure the efficiency of just one step in the process: how much of the energy in the electron beam can be turned into laser light. Extraction efficiencies ignore the cost of creating the electron beam in the first place. So when amazing new laser breakthroughs are announced, the key number to watch for space propulsion is the wall plug efficiency.

Earth-to-orbit shuttles will need several hundred megawatts of power, perhaps even a gigawatt. The raw power is available from existing power grids. The missing elements are a huge laser installation and the technology to keep the laser beam pointed correctly at the rising spaceships. Both missing elements also happen to be key goals of SDI.

The Army plans to build a series of large SDI lasers at the White Sands Missile Range in New Mexico. By 1991, a free electron laser with power in the tens of megawatts will be ready. The Army hopes to put a bigger laser operating at hundreds of megawatts into service by 1994. (The exact power levels are classified.) The laser, which may be strong enough to destroy Soviet satellites in orbit, will cost $1.5 billion to $2 billion.

The high-power FEL will be more than a mile long, including a 2,400-foot electron accelerator and a 660-foot magnetic "wiggler" to convert electron energy into photons. For SDI, ground-based lasers would flash their energy up to relay mirrors in high orbits and then down to "fighting mirrors" in low orbits. The mirrors are cheap, compared to the cost of a laser battle station in space. Many mirrors could be launched to frustrate any Soviet first strike against orbital defenses.

The ground-based FELs would be heavily shielded against Soviet laser attack. Because FELs are "survivable," the military is pouring enormous money into their development. The research blitz almost guarantees major laser improvements that can be used for space propulsion.

Creating the laser isn't the only problem. Aiming the lasers and mirrors at fast-moving targets (such as the laser-boosted engines of spacecraft) is a major hurdle. The SDI Organization has the same "pointing and tracking" problem in shooting Soviet missiles, and is spending billions of dollars to solve it.

The SDI work won't solve all the propulsion problems, unfortunately. Laser beams have a tendency to spread out when sent thousands of miles through space. One SDI test beam started a half-inch wide in Hawaii and spread to a quarter-mile wide by the time it reached the space shuttle, only 200 miles up. SDI researchers will try to control this spread, but they don't need to create a pinpoint focus. They can destroy a Soviet ICBM by broadly heating one side. During the stress of the boost phase, any uneven heating will cause expansion that ruptures the missile. The laser doesn't have to burn through the missile in a tight beam—but a tight beam is essential for propulsion. It's a problem that commercial firms may have to solve for themselves.

SDI researchers also wonder what happens to the atmosphere when gigawatt lasers are fired. They guess tremendous noise, artificial lightning, and ozone disturbances are likely. These side effects are acceptable when trying to prevent World War III but may be too destructive for a routine transportation system.

But military research will solve the core problem of getting money to build huge laser installations. NASA won't have to beg for the billions needed to construct laser sites. SDI also will make tremendous advances in technology to focus laser beams and track space targets. Unless SDI is derailed, it could soon provide the world with the building blocks needed for truly low-cost space transportation.

LASERS OPEN THE SOLAR SYSTEM

Reaching Mars or the asteroids using today's technology would be very expensive. Rocket fuel has to be hauled up the Earth's gravity well at great expense. Because fuel is so costly, travelers to Mars would have to use it efficiently in long, slow looping orbits. That, in turn, requires very expensive spaceships able to recycle air, water, and food. These "closed-cycle" ships would be away for two or three years, so every system would have to be foolproof. Massive costs for designing "perfect" equipment would be multiplied by the need to shave weight on every system at the same time.

Laser systems change the whole picture because the fuel doesn't have to be lifted off the Earth. The basic electric energy can be generated by ground units or solar power satellites in orbit. The fuel to be heated by the lasers can come from the Moon. Liquid oxygen would be a good choice for fuel because Moon soil is about 40 percent oxygen. The Moon can even supply basic building materials such as aluminum or plastic composites for the solar power satellites. (Details are available in Chapter 11.)

Cheap electricity from solar power satellites and cheap fuel from the Moon will let solar explorers plot direct routes to Mars and the asteroids. Round-trips to Mars could take three months instead of three years. Mars ships could use the partial recycling systems developed for space stations, skipping the expensive process of 100 percent recycling. A complete standby air and water system could be carried along, avoiding the expense of making the primary system foolproof.

The shuttle-dominated NASA budget provides little support for lasers and other advanced propulsion ideas. Before the *Challenger* accident, shuttle cost overruns soaked up every spare dollar. After the accident, the cost to fix the shuttle system and pay for a fourth orbiter kept most nonshuttle programs fighting for their lives.

The Marshall Space Flight Center near Huntsville, Alabama, is NASA's main center for propulsion research. Scientists there receive only $100,000 a year to work on lasers. They use an Army-surplus laser, capable of only 30 kilowatts. On their budget, they can't even attend SDI conferences, much less actively work with the military to make SDI lasers pay off for civilians. A technology that could give humanity open access to the entire solar system gets less federal support than some individual farmers get in subsidies. (Thousands of cotton and rice farmers received federal checks for $250,000 *each* to "promote exports" in 1987.) The misplaced investment is tragic.

The Marshall researchers are using hydrogen as the target for their laser. Hydrogen may be the ideal fuel because of its low molecular weight. It also can be a cooling fluid for the engines, which generate fantastic heat levels. Cold liquid hydrogen on its way to the laser chamber would first bathe the engine walls to keep them from melting.

Hydrogen gas has a drawback, however. It's transparent to laser light, so the beam just passes through instead of being absorbed. One way to catch the laser energy is to create a hydrogen plasma. The plasma is opaque and absorbs the laser energy. The Marshall researchers say hydrogen heated to a plasma of 20,000 degrees Kelvin (about 36,000 degrees Fahrenheit) can be the central laser-catching section of the engine. Fresh hydrogen gas will enter the engine and flow around the central plasma. The plasma heats the gas to 5,000 degrees Kelvin or even higher, perhaps up to 15,000 degrees. This produces explosive expansion and thrust.

Predictions for the hydrogen/laser combination show thrust four to eight times higher than engines working with the combustion of hydrogen and oxygen. Fuel tanks, instead of being taller than an office building, can be more like the size of a two-car garage. The cost of building the rockets would drop correspondingly.

A step beyond laser-boosted rockets is the laser-boosted spaceplane. The laser spaceplane wouldn't carry fuel at all. Instead, the beam would hit air coming into the engine. Explosive heating of the air would provide the thrust for propulsion.

Dr. Leik Myrabo of the Rensselaer Polytechnic Institute expects laser spaceplanes to be ten times more efficient than laser rockets. Despite this advantage, their creation will lag behind the first laser rockets. Laser spaceplanes will be much costlier to design because the ignition of thin air will be very tricky.

Dr. Myrabo and Dean Ing describe laser rockets and spaceplanes in a 1985 book called *The Future of Flight (Baen Publishing)*. Some space executives believe the book is too optimistic, but it does provide a visionary look at advanced transportation.

Laser-boosted ships are a good bet to be the first revolutionary transportation system. Other ideas include space elevators and anti-matter propulsion. These seem harder to achieve, but the true winner will depend on breakthroughs in technology over the next 20 years. Forecasting inventions for the next two decades is impossible, so any of these ideas could be the eventual winner.

ELEVATORS IN SPACE

The cheapest way to reach space might be an elevator driven by a standard electrical motor. A space elevator actually is possible, despite how weird it sounds.

An orbiting geostationary platform would drop a cable (called a *tether* in space applications) down toward the Earth. Because the platform is in geostationary orbit above a single point on the equator, the tether would simply hang down to a convenient mountain top. Once built, the space elevator could put people and cargo into orbit for just about the cost of electricity — under a dollar per pound.

The space elevator's big problem is finding a strong tether. A cable stretching 22,300 miles from orbit to the ground would snap of its own weight, using materials available today. Despite this problem, space experts believe it possible that future materials may be strong enough to work.

Space experts are very enthusiastic about tethers because even short ones would be useful. Right now, cables 12 to 200 miles long are being readied for shuttle experiments.

A 200-mile tether can help send payloads up to higher orbits by working somewhat like a slingshot. Imagine the shuttle reeling out a satellite on a tether, "up" and away from Earth. As the satellite moves into a higher orbit, it will tend to lag behind the shuttle, because the satellite must cover more distance in the higher orbit.

The tether prevents any lag, pulling the satellite forward and increasing its speed. At the same time, the tether drags on the shuttle, slowing it down slightly and dropping it to a lower orbit. Energy is transferred from the shuttle to the satellite as it reels out to higher and higher orbits. Because the shuttle is massive compared to the satellite, the shuttle slows slightly whereas the satellite speeds up a great deal.

After the satellite is reeled out to 200 miles, it's released. The extra speed given up by the shuttle flings the satellite outward to reach a higher orbit. The increase in the altitude is roughly seven times the length of the tether. The shuttle moves into a lower orbit, which actually is an advantage. The shuttle must return to Earth anyway. The lower orbit lets it burn less fuel when it brakes for reentry. With a long enough tether, the shuttle wouldn't need to burn any fuel to reenter. It simply would reel out a satellite until the shuttle's orbit dipped into the fringe of the atmosphere. Friction from the air then would bring the shuttle all the way down to reentry speed.

NASA and the Italian space agency are eager to test tether applications in a series of shuttle-based experiments. The first will attempt a tether 12 miles long, and the second will try for a tether 60 miles long. The 60-mile version will drop a payload "down" toward the Earth from the shuttle. The payload will be a scientific package to study the upper atmosphere. The shuttle will drag the instruments through air that's too thin to support balloon research but too thick to allow a free-flying satellite to stay aloft.

Moving from this simple tethered research probe to a system that can boost satellites and slow down shuttle orbiters is still under discussion. Ivan Bekey, NASA's director of advanced programs, is the leading U.S. planner. He calculates that satellites headed from the shuttle up to geostationary orbit could be 60 percent heavier and more powerful if they used a tether boost. If a geostationary platform dropped a tether down to meet the satellite, the gains would be doubled.

The strongest materials available for tethers now are man-made, such as DuPont's Kevlar and Allied's Spectra 1000. A Kevlar cable in low Earth orbit can be about 360 miles long before its own weight gets in the way.

Kevlar tethers can be longer than 360 miles if they are in higher orbits. The weaker gravity at geostationary orbit allows a tether almost 5,000 miles long.

The low gravity near the Moon would allow a 600-mile Kevlar cable from an orbiting platform. A tether extending "up" from the platform would catch incoming traffic. The incoming payload would transfer its speed and energy to the lunar platform, boosting it to a higher orbit. Then the cargo would be reeled out again, this time down toward the Moon. The second step slows the cargo still more and adds more energy to the lunar platform.

Outbound cargoes would reverse the process and drain energy from the lunar platform. A dangling tether would catch payloads that blast off from the Moon with a very small rocket. They'd be reeled in and then reeled out to a higher orbit. The lunar platform would drop to a lower orbit as its speed is transferred to the outgoing cargo. The whole cycle stores energy from incoming cargo to be given to outbound cargo. Paul Penzo of the Jet

Propulsion Laboratory estimates that a Moon tether would save 50 percent on the propellant needed for lunar landings and launches.

A similar system is possible at Mars. It has two moons, Phobos and Deimos, with close-in and nearly circular orbits. They would be way stations like the lunar platform with tethers dangling up and down. Almost like Tarzan, cargo carriers from Mars would grab the tether hanging down from Phobos, climb to the tip of the upward tether, and release. They'd swing outward to catch the tether hanging down from Deimos. After climbing the Deimos tether, they'd release and swing out for the trip to Earth.

The massive weight of Phobos and Deimos would keep them in stable orbits regardless of the cargoes coming in and out—the flows wouldn't have to be balanced. Heavier outbound traffic could "steal" energy from the Martian moons repeatedly and not affect their orbits.

Penzo calculates the Mars tether system would save a third of the propellant needed to travel between Mars and Earth. The tether system also would be ideal to use with a permanently orbiting Earth-Mars transfer ship. The transfer ship would cycle between the two planets, never stopping or even slowing down. All the living quarters, recycling equipment, and other long-term needs would be accelerated just once, when the transfer ship is launched. Small space taxis would meet the transfer ship as it passes each planet. Energy would be spent to accelerate and slow just the passengers and cargo docking with the transfer ship, rather than the entire interplanetary spacecraft. The Mars tether system would be able to throw cargoes to the transfer ship with only minuscule use of chemical rockets.

How Up and Down Exist in Space

First-time students of tethers often are confused by all the references to "up" and "down" in what's supposed to be the weightless environment of space. Orbital space near planets, however, does not have true zero gravity. What exists is simply a balance between gravity pulling an object down and centrifugal force flinging it outward.

Zero gravity occurs only at the exact center of an orbiting space structure. Only the center is traveling at a speed that exactly balances gravity with centrifugal force. A space station thus will have one "sweet spot." A few yards closer to Earth, gravity will start growing stronger than the centrifugal force. A few yards higher than the sweet spot, centrifugal force will be stronger than gravity. These forces will tend to pull a space station into an up-down alignment with Earth. One end of the station will be grabbed by gravity, and the other will be tugged away from Earth by centrifugal force. The same forces apply to tethers, pulling them either down to Earth or up away from their anchor in space.

Tethers for the Moon, Mars, and space stations can be built right now, from existing Kevlar cables. But the revolutionary Earth-to-orbit elevator must wait for new, stronger materials that won't snap of their own weight, perhaps sometime in the twenty-first century. Orbital research labs may create an ultra-thin wire made from a single crystal—a flexible diamond thousands of miles long—that does the job.

MAGNETIC LAUNCHERS

Another way to use electricity instead of rocket fuel involves magnetic launchers. A series of electromagnets along a tube can pull a payload forward, forcing it to blinding speed after only a few feet. The launcher spits out the cargo with a push several hundred times the force of gravity. The crushing acceleration means magnetic launchers are cargo devices that won't be used to fling humans into space.

Magnetic launchers might get their first use at Moon mining camps. They would hurl the mining camp's products back toward Earth for use by orbiting factories. As explained in Chapter 11, the Moon could be a source of aluminum, iron, oxygen and glass. Oxygen will be sorely needed for fuel and life support, while the other materials can be used to build space stations, solar power satellites, and ships to explore the solar system. Space economics is the reason the Moon will become a supply station: It takes 20 times more energy to bring a pound of building material up to orbit from the Earth than down to orbit from the Moon.

The Space Studies Institute at Princeton pioneered the civilian development of magnetic launchers. The institute calls them mass drivers. Prototype mass drivers at Princeton have shot tiny samples with a force of 1,500 gravities. That's just shy of the 1,800 gravities needed to hurl cargo to escape speed from the Moon. The institute calculates that a mass driver on the Moon would be only 100 yards long. Construction of this simple machine should be possible rather soon after we return to the Moon—the launcher is just a collection of magnets and some type of power supply. Yet it could be the essential link to supply orbital colonies with low-cost oxygen and building supplies.

The military began putting cash into mass driver research around 1980 for potential missile defense and for protection of Navy ships. The type of mass driver developed by the Space Studies Institute is getting some Pentagon support, because it's easy to build and operate. But the military favors a more powerful type, called a railgun. Railguns have two metal rails carrying a current and an armature that completes the circuit between the two rails. The armature's magnetic field pushes against the magnetic field of the rails, forcing the armature and its payload forward. (In electric motors, the armature is the spinning part. Railguns essentially are electric motors that accelerate the armature along a straight path rather than spinning it on a shaft.)

Railguns pack a tremendous punch. A Westinghouse railgun installed at an Army research center in Dover, New Jersey, launched a ten-ounce projectile at 4.2 kilometers per second. The Army now is looking for contractors to build accelerators capable of 10 to 100 kilometers per second. It's also looking for companies to build projectiles with electronic guidance systems that can survive more than 100,000 gravities of acceleration. This research is to come together by the early 1990s in Project Sagittar, when the military launches an orbiting battlestar to demonstrate the railgun.

Plans for battlestars may spur commercial space development. An orbiting battle station needs to be massive to fend off laser attacks, so the military may buy tons of Moon rock from private haulers. Enemies would be forced to spend precious minutes trying to burn through the shield of Moon rock to reach the satellite's weapons. An enemy couldn't launch a surprise first strike that takes out the orbiting defenses.

In addition, there may be spin-offs from railgun research that benefit the slower, simpler mass drivers to be used by private industry.

THE ULTIMATE ENGINE

Fans of "Star Trek" know that in the twenty-third century, the *Enterprise* is powered by anti-matter engines. Anti-matter is the mirror opposite of normal matter and would be the ultimate spaceship fuel. When brought together, an anti-proton and normal proton completely vaporize one another, turning all their mass directly into energy. Even a hydrogen bomb can't do that — the bomb only converts 1 percent of its mass to energy.

What may surprise Trekkies is the serious research now under way on anti-matter drives. The Air Force has studied anti-matter engines in-house and through contracts. Capt. Bill Sowell of the Rocket Propulsion Laboratory calculates that one gram of anti-protons has energy equal to 6,000 tons of rocket fuel. One gram is less than the weight of a dime.

Turning that energy into propulsion will suffer some waste, but Sowell estimates that perhaps half could be converted to thrust. That would make a dime's weight of anti-matter equal all the fuel used in a shuttle launch. What's more, all that power would be devoted to moving payload, rather than payload and enormous fuel tanks.

The hard part is actually getting some anti-matter. None has been found to exist naturally. Scientists suspect all of it was destroyed at the dawn of the universe when it reacted with normal matter. They speculate that normal matter slightly outnumbered anti-matter, making normal matter the sole survivor. Thus, anti-matter to fuel the ultimate engine must be created artificially.

Today's anti-matter factories are giant synchrotrons. The seven kilometer CERN (formerly, European Centre for Nuclear Research, now European Organization for Nuclear Research) synchrotron near Geneva, Switzerland, produces anti-protons by firing a particle beam at a tungsten metal target. CERN has produced a trillion anti-protons at a time with this technique. Unfortunately, it costs $40,000 in electricity and the trillion anti-protons equal the energy in a single shot from a child's cap gun.

Nothing like a gram of anti-matter, or even a milligram, has ever been created by man. But Sowell notes that the world production rate for anti-protons has increased by an order of magnitude — that's ten times the previous level — every two and a half years. Rocket Propulsion Lab officials hope to establish a factory by the year 2010 that produces a gram of anti-protons annually.

A synchrotron is a racetrack for atomic particles — it uses tremendous magnetic fields to accelerate and guide electrons and protons. Before 1987, synchrotrons required enormously expensive superconductors cooled to near -400 degrees Fahrenheit by liquid helium. Keeping the superconducting magnets that cold was very costly. Now revolutionary breakthroughs in superconductivity are stunning the scientific world — discoveries worthy of the Nobel Prize are announced almost every week. Instead of -400 degrees, superconductors now function at -140 degrees. Super magnets for particle accelerators can be cooled to -140 by liquid nitrogen, at a fraction of the former cost of liquid hydrogen. And dozens of research teams are racing to produce room temperature super magnets.

This science bonanza may lead to pilot plants making experimental amounts of anti-matter in only a decade. Creating large-scale factories still will require oceans of electric power.

Futurists believe they know how mankind will get the abundant cheap electricity needed for anti-matter factories. Cheap electricity will come from a combination of self-replicating robots and solar power. The robots would be sent into space to build huge solar power stations. In one scenario, a "family" of perhaps ten different robot models might be sent to the Moon or Mercury. Some robots would be designed to process raw materials, some would stamp out parts, and others would assemble the parts into duplicates of the original family.

Working together, they would use local materials to create hundreds of children, who each would create hundreds of additional robots. After a few years, there could be a million self-replicating robots ready for the next step: creating worker robots. These workers could turn vast areas of the Moon or Mercury into huge solar energy farms, using only local materials, or they could be sent into orbit to build solar power satellites.

An article by Dr. James R. Powell and Dr. Charles R. Pellegrino in the September 1986 issue of *Analog* estimates that turning Mercury into a solar power station could provide 50,000 times the entire energy used by the United States—all for the cost of the original family of robots and a little patience while they work.

The solar power from Mercury could be delivered by laser to Earth for everyday use and to orbiting anti-matter factories. (Anti-matter factories are the ultimate "locally unwanted land use"—if one blows blow up, it's a hundred times worse than a hydrogen bomb. So factories will be put into extremely distant orbits and operated by remote control.) Even an inefficient anti-matter factory could produce abundant fuel with the massive output of Mercury available at virtually no cost. The solar system would shrink dramatically because the tremendous power of anti-matter engines would make the planets only a few weeks apart. Starships, perhaps even one named the *Enterprise*, could start seeking out other solar systems.

All this hinges on producing self-replicating robots to build the power stations required for anti-matter factories. The robots would have to walk or roll around the Moon or Mercury by themselves, so they'd need vision systems and programming for self-controlled navigation. They also would need brains that could be built from simple local materials.

The vision systems already exist in crude forms. A robot Army truck has been tooling around a Martin Marietta test site near Colorado Springs. This "autonomous land vehicle" managed a speed of 10 kilometers per hour in the summer of 1986 over 4.4 kilometers of straight and curved roads.

DARPA hopes the robot truck can be upgraded soon to leave the road and make its way cross-country. A mobile robot like that on the Moon could search out mineral deposits or actually participate in the work of a mining camp. DARPA's work ensures that vision systems needed for Moon-mining robots will be available by the turn of the century. Truck-like robots could build solar farms, refine metals, and produce oxygen for fuel at zero labor cost.

The next step in robotics would be self-duplication, perhaps a few decades later, which means the first "family" of self-replicating robots could be on their way to Mercury by 2020 or 2030. Today's college students may live to see the first team of robotic settlers, and today's toddlers may reap the first waves of free power beamed back from Mercury in 2050. For them, the entire solar system will be their local neighborhood because of the tremendous power at their command. True adventurers among them will set out for nearby stars like Alpha Centauri, a five-year round-trip for anti-matter or laser-boosted engines.

WHY THE YEAR 2050 IS IMPORTANT TODAY

Few space activities will make a profit at today's shuttle launch costs of $6,000 per pound. But investors in any field never look solely at today's costs and conditions. They must look ahead to the next year and the next decade. That's why the expected drop in launch costs is so important today. Investors and corporations will see that small projects started now will reap large rewards when launch prices decline.

The situation is similar to land speculation. An investor may buy a farm that's losing money or only breaking even, if he believes that a new interstate highway or major industrial plant in the next few years would raise the farm's value. The same outlook is reasonable for a corporation building a pilot space factory for new electronic materials. The company will gain valuable experience and develop key technologies. When the new interstate arrives — in the form of lower launch costs — that corporation will be in position to expand production rapidly.

Space development thus gets a double boost from every drop in launch costs. Immediate opportunities are seized, and companies start making long-term investments to capitalize on the next round of reductions. A true frontier will develop: bustling activity where profits can be made and a fringe of explorers, prospectors, and adventurers probing new areas for the next gold rush. The explorers may seek something physical, perhaps mining the asteroids for strategic metals, or they may seek knowledge, perhaps from research in zero-gravity laboratories.

The $5,000 ticket could go on sale early in the twenty-first century or sometime later. It depends on the decisions and investments made now. NASA can play an important role in speeding space development. It needs more money for spaceplane research so it can build a prototype. After the prototype flies, private operators should build and operate the actual fleet. NASA would then begin work on the prototype of a laser-boosted ship. The government's commitment to early investment in each generation of launchers would give pioneers faith that new systems will emerge to make their own investments pay off.

The next chapters survey today's small band of explorers. Keep in mind that any list of space opportunities today is like a scouting report on the New World from Christopher Columbus. It's sketchy and probably will miss products and services that turn into billion-dollar industries. When the space stations and lower costs of the 1990s create a bigger flock of explorers, then we'll see the true outlines of the new space economy.

Space Communications: Ready for New Growth 5

The first corporate profits in space were made from communications satellites (comsats). In fact, the comsat era began with a corporate initiative in 1961. That's when American Telephone and Telegraph (AT&T) approached NASA with the idea of launching an experimental communications satellite, Telstar. AT&T handled construction of the satellite and repaid NASA for the cost of launching it.

Telstar lacked one key feature of today's comsats: It wasn't geostationary—it didn't stay fixed in one spot in the sky. Giant dish antennas controlled by computers tracked Telstar as it spun across the heavens. The Telstar orbit was too low to make it a "fixed" satellite.

Most comsats now are launched into a very high orbit 22,300 miles above the equator. This distance is the geosynchronous orbit—the forward travel of the satellite exactly matches the turning of the Earth. A ground antenna pointing up to the satellite always aims at the same spot in the sky. With Telstar's lower orbit, it moved forward more quickly than the Earth turned. It also didn't follow the equator; it looped up and down the latitudes just as shuttle missions do now.

Satellite owners in the 1970s made easy money. Demand for comsat services grew much faster than the supply. Companies wanting to offer comsat services couldn't just drop by Radio Shack, buy a comsat, and pop it into orbit. They first had to apply to the Federal Communications Commission (FCC) for permission and prove their new comsat was needed. This process took several years, followed by three or four years for the satellite to be custom-built by Ford Aerospace, Hughes, or RCA. Supply couldn't keep up with demand in the 1970s and early 1980s.

This easy street turned into a mean street a few years ago. The FCC began granting comsat licenses more freely, and new larger satellites carried many more channels. About 600 channels were available in 1987 on orbit for domestic U.S. service, but fewer than 400 were being used. (Each channel can carry hundreds of simultaneous phone conversations.) Excess capacity in 1988 continued to run in the 35 percent range. The glut has driven down rental rates for comsats, discouraging firms from building more comsats any time soon.

Even more discouraging for the comsat business is a new technology called fiber optics. Laid between two high-traffic points, a fiber optic cable can deliver better voice quality at lower cost than satellites. Comsat circuits fade out during rainstorms, and they have an annoying half-second lag that disrupts conversations. The half second is the time it takes the signal to travel 22,300 miles up from the sender and 22,300 miles back down to the receiver and for a response to make the return trip. People start talking simultaneously because the lag makes them think the other speaker has paused. Fiber cables go directly from point to point, so the lag is much shorter—even on a trans-Atlantic cable, the delay is only 0.04 seconds.

Compounding the comsat woes is the sorry state of rocket reliability. Firms can't be sure that a planned satellite actually will reach orbit and start generating cash to support the company. All the 1986 and 1987 launch failures—shuttle, Delta, Atlas, Titan, and Ariane combined—also forced insurance costs to triple. Insurance costs about 20 percent of a comsat's value but can escalate to 30 percent immediately after a rash of satellite failures.

The glut, the fiber threat, and the insurance muddle make comsats a business for gamblers. Experts are predicting the glut will be erased by 1991/92, but experts have been wrong before. As a result, only four new U.S. comsats have been put into production since 1982, and only a few are on order.

The comsat bust is frightening to corporations planning private launch services: Who's going to use their boosters? Comsats are far and away the largest group of customers for private rockets. Two thirds of the weight to be launched through the end of the century hinges on U.S. and international comsats, according to a study by Battelle Columbus Laboratories.

Demand for comsats could pick up again as existing satellites die from lack of fuel, exhausted electrical systems, and burned-out tubes. (Some comsats do, in fact, suffer from blown tubes just like old-fashioned TV sets. They use traveling wave tubes as amplifiers, and these fail much more often than, say, a home stereo system.)

Dying satellites will prune back the comsat glut. Otto Hoernig of American Satellite Company predicts on-orbit domestic channels will drop from 600 in 1987 to 400 in 1993 and 200 in 1994. Someday soon companies will be forced to start shopping for new comsats and boosters to launch them. The private rocket firms, however, need to find customers right now.

There are so few that only firms with military contracts are assured of staying in business.

DIRECT SATELLITE BROADCASTS TO HOMES

The major domestic use of comsats is to beam TV programs down to network affiliates and cable systems, into dish antennas the size of a garage. These same beams can be picked up by individuals equipped with six- to ten-foot dishes. It's an expensive hobby—the dish antennas and extremely sensitive tuners needed to catch the shower of TV programs from space can cost $2,000 to $10,000.

In the early 1980s, many entrepreneurs wanted to expand this market by using higher power comsats. The signals could be beamed to small dishes, perhaps only two or three feet wide, and processed by relatively inexpensive tuners. A system would cost about $500.

Many firms dreamed that direct broadcast satellites (DBSs) for television would make them rich—they would create instant national networks. Eight companies won permission from the FCC in 1983 to start construction of DBSs. Most of their dreams have been dashed against the hard rocks of competition and rival technologies.

Strange as it may seem now, in 1981 there were virtually no video cassette recorders (VCRs) in use. If you wanted to see a movie, you paid $4 for a ticket or waited a few years for the TV networks to show it. Or you paid for Home Box Office (HBO) and Showtime if you were in the lucky one third of households with a cable system available. Fewer than 10,000 people had backyard satellite dishes.

Today more than 28 million VCRs sit next to TV sets across the country. Six months after theatrical release, most films are available for a dollar or two rental fee from video shops in even the smallest villages. More than 1.5 million Americans have invested in backyard satellite dishes. And 75 percent of American homes have cable available to them.

Who needs direct broadcast now?

Remote rural families desperate for TV programming already have backyard dishes. Everyone else has cable service or expects to get it in a few years. And most TV-hungry homes are equipped with a VCR.

This technological revolution blitzed DBS from one side, while lack of good programs undermined from the other. The potential DBS operators discovered that finding good programs to broadcast actually was more important than winning a federal license to launch a satellite. Program suppliers like HBO and Cable News Network (CNN) were not eager to sell to DBS operators. They feared local cable companies would boycott any program supplier that sold to a DBS operator, because that would be a challenge to their local monopolies.

Program suppliers also realized the tremendous value of their movie libraries and news staffs. Companies like HBO and CNN demanded high fees that DBS systems weren't able to pay, or they demanded part ownership in the systems, which DBS operators weren't willing to grant.

Only a few companies now seem to be serious about building high-power DBS systems. One is TCI/Tempo, which plans a 32-channel system able to reach 1.5-foot dishes. In a 1988 speech, however, company officials admitted they hadn't yet lined up any of the crucial programming needed to make their service attractive to customers.

DBS COULD RISE AGAIN

The first crop of DBS firms were washed out by competing VCR and cable technologies, and by the tough stand of program suppliers. Both factors may turn around soon, to aid rather than thwart DBS systems.

The new technology that may save DBS is high-definition television (HDTV). HDTV promises a much sharper picture broadcast in a new wide format to match the shape of a movie screen. The sharper picture would be created by using more horizontal scan lines to paint the image. Current U.S. television is among the fuzziest in the world because only 525 scan lines are broadcast. The inferior picture has a simple cause—American technical standards were set by the pioneering equipment of the 1940s. European systems use 625 scan lines, because their standards were set later when TV equipment was more advanced.

Today's high-tech electronics could deliver pictures with double the sharpness by doubling the number of scan lines. But the new video faces a chicken-and-egg problem: Television stations won't convert to high-definition broadcasts until people buy high-definition sets, and people won't buy the sets until there's something to see. DBS could be the answer. A pair of direct-broadcast satellites could provide 16 high-definition channels for the entire country in one stroke. Production of HDTV sets would follow easily—firms like Sony, RCA, Zenith and Phillips would make millions by upgrading consumers to a more expensive video system.

New VCRs equipped to play HDTV tapes also would spread the new technology even before over-the-air broadcast stations offered HDTV programs.

The dazzling picture and digital sound quality of DBS would suddenly give it a competitive advantage over cable systems. This may prove very tempting to the movie industry. Paramount or Disney might decide they could wring more cash out of their movies by showing them first on their own high-definition DBS systems. Only after several months on DBS would movies be released to the VCR market, to be followed later by the cable market and even later by the TV networks.

The movie studios finally would have the power to strike back at HBO, which they resent for its near-monopoly power. The studios are helpless against HBO now. Its competitors like Showtime simply don't have very many customers, so studios that don't like the price offered by HBO don't have much of an alternative. They certainly can't afford to build their own duplicate cable systems to every household, even if they could get legal permission to do so. But the low cost of renting a few channels on a "condominium" direct-broadcast satellite will give them an affordable alternative to the cable networks.

The combination of a sharper picture and premium movie programming probably will push DBS into reality by 1991 or 1992. The long wait until 1991 isn't due to technical problems. High-definition equipment is already being used by production companies around the world. The delay is caused by European firms and politicians who fear a rapid endorsement of a world standard for HDTV. They worry that a blitz of Japanese sets and American satellites would bury their own manufacturers and ruin the government-run TV monopolies. They are playing for time, so European firms can learn how to produce high-definition equipment and government broadcasters can figure out ways to limit competition from outsiders.

Direct-broadcast satellites might bring about the rebirth of the comsat industry and with it a significant new market for private launch firms. Each DBS system requires two comsats to cover the nation adequately. And quite a few DBS systems could be created, because movie studios won't be the only players in the DBS field. Anyone seeking an instant national network would be a potential DBS operator. The British Broadcasting Corporation (BBC), for example, might instead set up an American DBS pay channel to reach the upper-crust audiences with the money to afford advance access to its programs. Public TV stations would get the BBC material a year or two later.

The ability of DBS to deliver a national audience for peanuts will create opportunities for special-interest broadcasters. Putting together a national network now is extremely difficult — one must convince hundreds of local stations to carry the program and then pay for expensive market research to prove to advertisers that someone is watching. Renting a national DBS channel will be one-stop shopping. And advertisers aren't required, because DBS can be operated as a pay service through scrambled signals.

If these predictions of wonderful new programs sound familiar, they are—cable systems promised the same type of new era. Cable did bring a few new things, such as 24-hour news, 24-hour sports and 24-hour preaching. But there is no *Field & Stream* channel, no *Omni* channel and no *Vogue* channel. Local cable systems generally won't carry a new service unless it's available full time. And the costs of producing and marketing a 24-hour nationwide *Vogue* channel, for example, would be prohibitive for most start-up firms.

With DBS, a narrow audience can be reached without a huge bankroll. A *Field & Stream* service could start with just a few hours of programming each month. The service would rent time on a DBS channel for those hours and send the programs out scrambled. Only the DBS homes that paid the decoding fee for that *Field & Stream* segment would get to see it. Nationally, there is no way for existing cable systems to offer the same pay-per-view system to small programmers. You either supply a complete 24-hour service with wide appeal, or you don't get on most cable systems.

Television suddenly could have the same diversity as magazines, with special services for hundreds of groups. For the 18 million Americans who don't have English as their mother tongue, there will be Arabic networks, Asian networks, and additional Hispanic networks. (Univision, formerly called Spanish International Network, now dominates the Hispanic market through its 409 standard broadcast TV station affiliates.) An Asian network is especially suited for high-definition DBS TV because HDTV technical standards will allow four to six sound channels. The video for one Asian movie or news show could be accompanied by four different languages on the sound channels. Programmers could send out the sound track in some combination of Japanese, Mandarin Chinese, Canton Chinese, Vietnamese, Cambodian, and Laotian. Multilingual sound channels in the HDTV standard also will permit European broadcasts to reach larger audiences. European firms want to broadcast simultaneously in various combinations of English, French, German, Italian, Spanish, Portuguese, Dutch, Swedish, Norwegian, and Flemish.

Once high-definition DBS systems get going, it's clear they could spawn a new growth industry in broadcasting. And they would provide private booster firms with something to launch.

THE GEOSTAR REVOLUTION

Another new space-based industry is on the way, one that merges go-anywhere communications with a position-tracking service. Busy people will have a constant link with the world communications network, whether they are riding in a taxi or visiting a client's office. The same service will allow companies to monitor—from space—the movements of any valuable item, from trucks to motorcycle couriers.

The new service is the invention of Dr. Gerard K. O'Neill. He conceived the idea while a physics professor at Princeton University, and later founded the Geostar Corporation to exploit it. Geostar customers will use hand-held transceivers to send and receive messages relayed from two satellites in geostationary orbit. The satellites will link the transceivers with a central supercomputer that serves as the switchboard for messages.

Users would see incoming messages on a small computer screen and tap out replies on a small keypad. Units mounted in trucks, trains and boats could be larger, if only to provide a full-size keyboard. Sony and Hughes are building $3,000 first-generation terminals now for vehicle use; the personal unit may sell for around $500 to $1,000. The Geostar central supercomputer also will track the precise location of each transceiver.

Position-tracking is the key Geostar innovation. Fleet owners will know the exact locations of their moving vans, semitrailers, and railroad boxcars anywhere in the country. Police departments will know precisely which squad cars are closest to reported crimes. And citizens who carry Geostar units will be able to call police instantly to their position when they are threatened by crime in the street or in their homes. Potential uses are enormous, and Geostar is likely to hit a billion dollars in sales in six or seven years.

The central computer will calculate the user's location by timing how long the transceiver signal takes to reach each of the two satellites. Geostar Central can send the location data back to the original transceiver, or forward it to another site. Consider Mayflower Transport—headquarters would love to know exactly where all its moving vans are. Drivers who goof off with a three-hour lunch stop couldn't blame their late arrival on traffic. And customers could be told rather precisely when to expect their furniture. Mayflower has signed up to buy $2 million of Geostar terminals and $400,000 a year in Geostar services.

Factories operating with a "just-in-time" inventory system would be ideal for Geostar services—managers could command trucks in transit just like air traffic controllers. Trucks arriving too soon would be ordered to slow down, and overdue truckers might be told to skip lunch to speed their arrival.

Security would be vastly improved. Hijacked trucks would set off alarms at the home office the moment they left the planned route, or the second the cargo doors were pried open.

Safety also could improve for people living near busy truck and rail routes. Any trucks or trains carrying hazardous cargoes could be required to carry Geostar units, programmed with the nature of the hazardous materials. Local police or highway patrol headquarters could monitor these shipments, using a personal computer that displays the trucks moving across the screen. The software would sound an alarm for any hazardous truck breaking the speed limit, leaving an approved route, or becoming involved in an accident. A monitoring system could ensure that hazardous waste leaves a factory and goes directly to an approved disposal site, rather than to an illegal dump.

The space monitoring also works with maritime shipments. For example, the Coast Guard requires that ocean dumping take place at least 50 miles off shore. Officers doubt that all barges really go the full distance. A Geostar monitoring system would guarantee compliance.

The list of Geostar applications seems endless:

Overcrowded prisons may be thinned by putting petty criminals under house arrest. A Geostar collar would track their movements. They or their families would pay for food and board, taking the burden off taxpayers.

A Geostar collar also would be dandy for paroled criminals – they could be required to go straight to work and back. One step into the park to buy drugs or into the subway to molest people would send them back to jail.

How Geostar Aids Consumers

Geostar may create new kinds of taxi services, from economy class to premium class. The new cut-rate services would use multipassenger jitneys dispatched by a central office. As customers signaled for pickups via Geostar, the central office would assign them to the nearest jitney going the right direction. Their pickup request could be confirmed almost instantly, with the estimated time of arrival for the jitney, and the estimated time they'd arrive at their destination.

Premium taxi services could operate much the same way, substituting a luxury one-fare cab for the shared-trip jitney. No premium services operate now because cab consumers simply take the first taxi that shows up. Cab companies have no incentive to upgrade their fleets with air conditioning, more legroom, or bonded drivers who are guaranteed to drive safely. Extra efforts don't result in any extra passengers.

With Geostar, upscale consumers would have a convenient way to summon special first-class service. Only a small premium fleet would be needed, because one premium cab sitting parked could "cover" several city blocks via Geostar. A premium fleet thus could respond quickly without requiring a massive investment in cars. Passengers could signal for a pickup from inside a store or office building, instead of standing outside in the rain.

Geostar also will be a hot accessory to install in private cars. It will help drivers navigate and protect against theft. Geostar-equipped cars will have a dashboard map display, showing the exact location and direction of the car at all times. Some units will be able to speak, giving directions at each intersection. Thieves will avoid Geostar cars because the cars will signal police the instant they are stolen. (Authorized drivers will enter a personal identification number, similar to a bank cash card, to bypass the alarm.)

If a thief does pirate a Geostar car to the local chop shop (where cars are cut up for sale as replacement parts), police will get constant updates on the car's position. Stealing a car will become next to impossible.

The same theft detection is possible for any valuable object, and for children as well, perhaps with basic units tucked into their underwear. Geostar Central would watch to ensure the child stays within a preset area, like a yard or park, or within a few yards of the parent's transceiver. Child snatching would become much more difficult.

How to Start a Revolutionary Company from Scratch

The Geostar idea was born in 1978, following airline disasters in San Diego and the Canary Islands that killed hundreds of people. Appalled by the carnage, Dr. O'Neill decided that satellite technology and modern computers could be combined to make air travel safer. He filed for a patent on his new position-tracking scheme in the fall of 1980.

The patent was approved in 1982, and it recognized 49 separate inventions. What Dr. O'Neill couldn't get recognized, however, was the need for a new air traffic control system. The Federal Aviation Administration (FAA) just wasn't interested.

Many pilots predicted the FAA would never adopt Dr. O'Neill's invention and urged him to market the system for general use. They thought that wide commercial success would eventually force the FAA to use it for air traffic control.

Dr. O'Neill incorporated Geostar in February 1983 and began raising money. Most budding businessmen hit a roadblock at this crucial point, but not Dr. O'Neill: "I raised the first few hundred thousand dollars one afternoon by phone calls to about a dozen people," he recalls.

The quick cash flowed from Dr. O'Neill's fame as a space visionary. His book, *The High Frontier: Human Colonies in Space*, earned him respect as a space guru. He also founded the Space Studies Institute, a nonprofit foundation supported by hard-core space enthusiasts who make five-year pledges of $500 or more. This steady income lets the institute fund pioneering research, such as chemical and mechanical ways to turn lunar soil into building materials, and development of mass drivers to launch those materials back to low Earth orbit economically.

His fame also allowed Dr. O'Neill to attract an all-star board of directors. That made investors feel comfortable giving their savings to a start-up firm in a start-up industry. One early board member was Dr. Luis Alvarez, a Nobel Prize winner in physics who also invented the ground-controlled approach method still used by Air Force planes.

The author of this book invested in Geostar in 1986, and for a time was an executive of a Geostar subsidiary.

The early investors believed that $300 million would be required to build the Geostar system, a figure far beyond the $10 million to $50 million that a new company can hope to raise from stock sales.

Dr. O'Neill and his allies devised an ingenious game plan to turn a tiny amount of cash into a revolutionary communications system. The key plays set up strategic partnerships with major corporations. Some companies put up cash to get exclusive franchises in certain areas; other firms traded their services for Geostar stock.

The most important barter deal involved Comsat, the company chartered by Congress to handle all U.S. satellite traffic with the worldwide Intelsat consortium. Comsat Technology Products agreed to build Geostar's central computer system and write its software in return for Geostar stock. Comsat's contribution is valued at $2.8 million—but the Comsat endorsement probably was worth millions more in coaxing money from skeptical investors.

A railroad company, Guilford Transportation Industries, made another key play. Guilford took an exclusive license to provide Geostar to railroad users through its Railstar subsidiary. It agreed to purchase $5 million of Geostar stock in regular installments as Geostar completed each phase of its start-up. The agreement lent credibility to Geostar's market projections—a potential customer was putting up millions to finance the system.

Dr. O'Neill and Martin Rothblatt, the president of Geostar, also pulled off a coup with the mobile user terminals. Almost all the engineering and prototyping costs are being paid by other companies. Sony and Hughes agreed to build the prototype units to be carried by trucks and trains.

The initial transceivers are expensive, at $3,000 apiece, and not very portable, at 25 pounds. The Geostar executives attacked this problem head-on—they persuaded the U.S. government to pay for shrinking the terminals down to pocket size. The U.S. Customs Service is funding the effort needed to get a 20-cubic-inch transceiver. The government will use the tiny units to keep contact with field agents from a number of agencies. The government-paid research will permit Geostar to bring out the advanced $500 consumer version of its transceiver.

Geostar also won important White House support, after Dr. Alvarez introduced Geostar executives to Dr. George Keyworth, President Reagan's first science adviser. He reviewed Geostar's plan and made crucial phone calls to the FCC when license issues were being decided. Keyworth told the FCC the White House strongly felt Geostar was in the public interest, Dr. O'Neill says.

Geostar executives also won aid from NASA. The space agency gave Geostar one of its first "fly now, pay later" contracts. The company gets to launch its satellites on the shuttle without putting up any cash. NASA agreed to wait two years after each launch before starting to bill Geostar. At that point, Geostar gets five years to repay the launch cost plus interest on the loan. The NASA deal—covering $65 million in launch costs—is called a "space services development agreement." NASA grants them to companies with significant new uses for space, after it evaluates the proposed business. NASA in effect floats a loan for the launch price, and shares the risk of a business failure.

Geostar executives also built strong ties with the European space community. They enlisted CNES, the French national space agency, to create a European Geostar system called Locstar. Geostar will get royalties from this system but Europeans will provide all the money to build it. The systems will be compatible; businessmen and tourists carrying transceivers from either system will be able to use them on both sides of the Atlantic.

Geostar is aiming for a global network. It's negotiated study agreements with the Indian space agency and the Australian satellite company Aussat on Geostar packages to serve the Indian and Pacific ocean regions. Executives would get their messages whether they were in Tokyo, Singapore, or Katmandu.

Even with all this wheeling and dealing, Geostar was still short of the hundreds of millions of dollars needed to construct its mass-market system. Geostar was in a bind until its leaders came up with a final series of trick plays.

The first inspiration came after field tests in the Sierra Nevada mountains south of Lake Tahoe in 1983. The field tests involved relays placed on mountain peaks to simulate the satellites. The system successfully guided pedestrians to hidden markers and small planes to landing strips. But as Dr. O'Neill extrapolated this experience to a full-up three satellite system, he saw the computations for a three-dimensional location fix would be very cumbersome.

The answer was to build a system with two satellites instead of three and just make two-dimensional computations. The change meant Geostar would not calculate a transceiver's altitude. This third dimension would be drawn from a digital terrain map of the United States—the Geostar computer would check the map for the elevation of all ground users. Airborne users would have to rely on an altimeter. The two-satellite system slashed costs for the space segment while making the ground computer calculations much easier and more accurate.

Then the Geostar executives realized they didn't need to start by launching their own dedicated satellites. They could create a junior Geostar by piggybacking small Geostar relays on other companies' comsats. That postponed the costly dedicated satellites until much later in the firm's growth. By hitchhiking on two comsats, Geostar brought its financing needs down from $300 million to only $80 million.

Finally, Geostar downsized its needs one more step. Executives planned a forerunner service that would use only a single relay. This provides one-way communications, from mobile units to headquarters. Mobile transceivers use an existing radio beacon system called Loran-C to get a rough fix on their positions. They send that information back to the home office along with brief messages or alarms (in the case of truck accidents, for example).

In March 1985, Geostar negotiated a deal with GTE Satellite Corporation to put the initial Geostar relay on the firm's GSTAR II comsat. A year later, GSTAR II went into orbit on an Ariane rocket, and Geostar was ready for business. Fleet owners had already signed up for 5,000 of the one-way units.

The appeal of Geostar for trucking firms shows up clearly on a software package called Fleetview created by Techwest of Richmond, British Columbia. Fleetview shows an animated map with all of a company's trucks, right on the screen of an IBM-AT class computer. Users can zoom in on particular regions to see which trucks might be arriving late. Users also can set up special "watch-for" conditions, and the software will flash to attention when those conditions are met. Mechanics or backup trucks can be dispatched the moment a truck breaks down or a refrigerated truck starts to overheat.

Disaster Number One

The GSTAR II comsat had two dangerous steps to complete after the Ariane launch. The first was successfully firing the "apogee kick" motor. Comsats usually are launched into egg-shaped orbits initially, with a low point several hundred miles above Earth and the apogee at the 22,300-mile geostationary orbit. The initial elliptical orbit is called a transfer orbit.

After a few circuits around the transfer orbit, the apogee kick motor fires when the satellite reaches the top height. This slams the satellite forward with enough energy to make the orbit circular at 22,300 miles. Misfires or no-fires of the apogee kick motor leave the satellite in a useless orbit. The kick motor worked.

The satellite's second hurdle was extending its solar panels and antennas. To fit into the rocket's nose during launch, the solar panels are folded back and strapped into place with a belt. To release the solar panels and get full electrical power, ground controllers must detonate explosive bolts to release the belt. To everyone's relief, GSTAR II unfolded its solar panels on schedule.

Geostar technicians conducted on-orbit tests of the package during April 1986. The primary signal relay and the backup relay met or exceeded all specifications.

Then both signal relays died.

One of the relays went completely dark, as if someone had pulled the power plug. Then two weeks later, the other relay went into a coma. Technicians could get tiny responses from the unit, but they were much too weak to carry commercial traffic.

To this day, engineers for Geostar, GTE, and RCA are puzzled by the dual failure. Investigators discovered that rocket fuel had been spilled on the satellite during the prelaunch preparations, but they saw no clear way for the spilled hydrazine to attack just Geostar parts and leave the rest of the comsat functioning perfectly. Their best guess is that power cables broke after chafing against some sharp edge. And engineers note that the rushed schedule for adding the Geostar relays to the GSTAR satellite meant the Geostar unit didn't get thorough on-the-ground testing.

The cost of the relays was partly insured, but there was a big problem: Satellite launches had come to a complete halt. The shuttle and most American rockets were grounded by disasters. Then the Ariane failed, too, on the mission right after GSTAR II's launch. Geostar faced at least a year's delay in starting the crucial initial service. Despite raising a tremendous $37 million from individual and institutional investors, by the fall executives were wondering whether they'd be forced to sell even more shares to stay afloat.

Rothblatt devised a brilliant comeback strategy early in 1987: market low-accuracy service derived from a government weather satellite. The polar-orbiting satellites carry a French radio receiver that was used to track slow-moving beacons tossed onto such things as ice floes, as an aid to forecasting. The Geostar engineers devised a way to use the same beacons on trucks, despite their far faster speeds.

Because the French device was put on the weather satellites to collect meteorological information, every Geostar truck beacon also carried a thermometer so it could send back temperature readings. That satisfied the government's desire to keep the satellite dedicated to meteorology. At the same time, it allowed Geostar to sell a commercial position-finding service without having any space assets of its own.

Because the weather satellites operate worldwide, Geostar now sells this service to ocean shipping lines as a way to track their ships globally.

Disaster Number Two

Dr. O'Neill had left daily Geostar management in June 1985 when he was hospitalized with a complication involving a rare form of leukemia. He recovered in November and began full-time work on the National Commission on Space.

After the commission issued its final report in March 1986, Dr. O'Neill tried to resume daily control of Geostar as its chief executive officer. The managers beneath him decided to force his retirement.

Dr. O'Neill, as a brilliant physics professor and space visionary, has a wide following. But his style as a business executive clashed with his top deputies at Geostar. That led to the June 1986 palace revolt.

In July, Geostar's board of directors approved a retirement package for Dr. O'Neill, who joined in the unanimous board vote for the package. His title would become "founder of the company" with the details of this new role to be negotiated. Lawyers for Dr. O'Neill and Geostar met repeatedly during July, August, and September to write a formal contract.

Finally, the attorneys set September 12 to settle on the contract and pay Dr. O'Neill accelerated royalties on his patent.

But on September 11 the Geostar managers discovered a hidden battle was under way for control of the company. Dr. O'Neill had been quietly collecting proxies from major stockholders so that he could remove Rothblatt and other key executives from the Geostar board. Dr. O'Neill thought their membership on the board was a conflict of interest — they sat on the same body that supervised them. Instead of company insiders, he wanted the Geostar board filled by nationally eminent people to increase the company's prestige and access to decision makers.

One victory for Geostar managers came in New Jersey Superior Court. The court blocked Dr. O'Neill from voting the stock he had given to the Space Studies Institute. Anti-O'Neill forces argued that the institute could lose its nonprofit status if its shares were voted in a proxy war. For the 375 investors who had poured their cash into Geostar, the O'Neill proxy fight and the loss of the first satellite made fall 1986 a gut-wrenching time.

The board of directors rallied around the management, and they replaced Dr. O'Neill as chairman with William Simon, a former U.S. Treasury secretary. Dr. O'Neill then agreed to keep out of proxy fights for five years. Investors stopped taking extra-strength tranquilizers and began counting their future profits again.

The next Geostar relay was shot into orbit in spring 1988 as part of a GSTAR III communications satellite launched on an Ariane rocket. The company began its System 2.0 service that summer. The initial users pay $45 a month for a subscription plus a small charge for each message sent. Geostar executives expect to become profitable by the end of 1990. An Arthur D. Little Company consulting study projects 1 million users for the advanced two-way service. Geostar profits could reach the $100 million level at that point.

THE COMPETITION

The Pentagon is creating its own set of navigation and position-fixing satellites that are fundamentally different from Geostar. The Pentagon's NAVSTAR system is one-way only: The satellites send out time signals from atomic clocks. User terminals will take the signals from several satellites and perform the calculations needed to determine their positions. The military prefers the NAVSTAR method because the user terminals are strictly receivers; they don't advertise their positions to the enemy by broadcasting to the satellites as the Geostar terminals do.

Armored columns on the move will plot their progress with NAVSTAR, as will fighter pilots and almost every other type of mobile military unit.

Even submarine-launched Trident D-5 missiles and land-launched MX missiles will use NAVSTAR in flight to improve their accuracy.

The NAVSTAR system puts the computing burden on the user terminal, however. The complex computer circuits needed to analyze the atomic clock signals from three or four satellites are very expensive. Even the basic receivers allowed on the civilian market cost about $3,000. These models are less accurate than the military units because they can't decipher the coded military signals needed for precise location fixes. Civilian NAVSTAR users will have to make do with 200-meter accuracy, compared with 6 to 10 meters for Geostar.

The NAVSTAR system now has seven satellites in a variety of 12,500-mile high orbits. They aren't fixed in geostationary orbits over the equator because that would leave big gaps in the polar regions that are crucial for strategic missiles and Navy fleets.

The Pentagon needs at least 18 satellites in orbit to get continuous global coverage. To reach that level, and to replace the early NAVSTAR satellites that only have five-year life spans, the Pentagon plans to launch 48 additional satellites through 1995. This huge demand makes the Pentagon a major customer for expendable launch vehicles.

NAVSTAR's timing signals have an important use in addition to navigation: They are key to secret government communications codes. The code used in a message — and the frequency on which it is broadcast — will change from second to second, so the receiver and transmitter need to be synchronized precisely. The cesium atomic clocks on the NAVSTAR satellites will be the link that keeps transmitters and receivers in tune with each other.

CIVILIAN COMPETITORS EYE THE MOBILE MARKET

Several firms want to win a share of Geostar's market, and are seeking customers with related services. One firm, Qualcom, has gained some success by combining Loran positioning with space relay on Ku-band satellites.

Another eight companies think Geostar's telegraphic service is too limited. They formed the American Mobile Satellite Consortium to offer a national mobile telephone service. Voice-oriented companies argue that consumers want the convenience of telephone-style contact; they don't think Geostar's tiny display screens and miniature keyboards are the best technology.

On the other hand, voice communications is much less efficient than

Geostar's service. Telephone conversations will take far more broadcasting power, and use up far more of the limited frequencies available in the radio spectrum. That means the convenience of a nationwide car-phone system may come with a stiff price tag.

Geostar's response has been to keep its transmissions in compact written form, but add devices at the user terminal to announce the messages in a computer-generated voice.

All these new services — position-finder satellites, a national car-phone system, direct-to-home TV broadcasts — may prop up demand for launch services. The private launch industry will need all the new demand it can get because the existing comsat industry will be reeling from its battle with fiber cable alternatives.

THE SPACE STUDIES INSTITUTE WILDCARD

If Geostar succeeds, it will create a new force in space development because Dr. O'Neill gave the Space Studies Institute 1.7 million shares of Geostar stock. By the early 1990s that hoard could give the institute an endowment of more than $100 million and an annual budget of $10 million to $15 million. The institute could become the first dependable source of major space funding outside of NASA and other governments.

The institute's budget still would be petty cash compared with NASA's annual $10 billion appropriation. But the institute is free of bureaucratic weight and dedicated to the sole purpose of creating an orbital economy based on lunar and asteroidal resources. Dr. O'Neill makes a comparison between the way Geostar and the government spend money. He says informed guesses place the cost of the completed government NAVSTAR system at $30 billion, whereas Geostar may cost about $300 million. That's a 100:1 efficiency advantage for private money over government money in developing similar space projects.

NAVSTAR does perform nonnavigation tasks such as coded communications and nuclear blast detection, so it's not directly comparable to Geostar. But the nonnavigation uses also underline how government programs usually can't stick to a single well-defined goal. Accounting for coded communications and nuclear blast detection, Geostar may have only a 50:1 cost advantage over NAVSTAR.

The impact of development money from the Space Studies Institute will go beyond the 50:1 efficiency. Dr. O'Neill hopes the institute's cash will draw in matching funds from private investors and the government. For example, the institute could develop a satellite to map the Moon for valuable mineral sites. The government might then donate the launch vehicle. Or it could develop the technology needed to gather up external tanks from the space shuttle and store them on orbit for later use as habitats and work areas. Private investors would be needed only at the final stage, when the project requires big money to start actual operations. The institute may be an incubator for many space projects that eventually are funded commercially for full development.

Wonder Drugs and Electronics from Orbital Factories   6

A popular T-shirt among scientists says:

GRAVITY
Not Just a Nice Idea —
It's the LAW!

And it's been the law for the entire history of the human race. Gravity has always been tugging at our work. When scientists mix chemicals, gravity pulls the heavier elements down to the bottom of their beakers. When they heat the beakers, the hotter and lighter parts always rise to the top.

Now we've discovered a new laboratory where weight is irrelevant. The space laboratory provides the revolutionary tools of *universal mixing* and *motionless heating*.

Universal mixing: Virtually any materials can be mixed together in the zero gravity of space, even oil and water, and they will stay mixed. Heavier particles don't fall to the bottom. Bubbles don't float to the surface.

Motionless heating: Heated materials stay calm. Fluids and gases don't churn like water does when it boils. There are no convection currents turning them into chaotic swirls. (Forces other than convection still work, so sometimes perfect stillness isn't possible.)

Mixing and heating are the most basic steps in science and manufacturing. Millions of experiments done under the rule of gravity now can be repeated for possible new results. Only a few dozen experiments have tested mixing and heating in space, so our ignorance about zero-gravity science and manufacturing is almost breathtaking.

One of the great hopes for space manufacturing is a metal or plastic alloy that would be a superconductor of electricity at room temperature. Superconductors pass electrical currents at almost 100 percent efficiency—no energy is turned into waste heat. Eliminating waste would save millions of dollars in long-distance energy transmission. New superconductors would lead to cheap yet extremely powerful electromagnets, a key part in everything from electric motors to medical imaging devices.

Today, metals can be made into superconductors only by cooling them down to almost absolute zero. Metals that superconduct at room temperature would be much more practical.

Scientists have drawn up a list of 245 potential alloys that might be superconductors at room temperature. Zinc and lead alloys are on the list, for example, but they haven't been tested. Most zinc-lead alloys can't be made on the ground because of weight differences. As molten zinc and lead start to solidify, the zinc rises to the surface because it's less dense than the lead. Thus, no one's explored various mixtures of lead and zinc to see if one might be a superconductor.

Scientists have been exploring other compounds for superconductivity and recently have made astounding progress. Exotic mixtures of oxides and ceramics are pushing the temperature limit on superconductivity from near absolute zero up to about -130 degrees Fahrenheit. A leader in these breakthroughs is Dr. Paul Chu of the University of Houston, who made his initial discoveries while leading a NASA-funded center in materials research.

This ground-based research bonanza may lessen the need for space work on superconductivity, or it may add urgency to space tests. That's because ground samples have been ceramics and oxides rather than metal alloys, and thus may be hard to manufacture into wire and other parts needed in practical applications. Space research could extend superconductivity into easily used metal alloys.

Europe's Intospace consortium believes microgravity will be essential in superconductor research. Intospace plans to launch its "Suleika" experiment in late 1989, and is seeking launch slots on the U.S. shuttle, the Soviet Photon, or the Chinese Long March rocket. The Suleika experiment will use containerless processing, employing acoustic levitation to create ultrapure materials.

The list of 245 possible *metal* superconductors targets one particular property. Besides superconductivity, there are a dozen other needs: strength, high melting point, reflectivity, corrosion resistance, magnetism, and so on. When inventors can start testing new alloys in space-based labs, breakthroughs are possible in all of these areas.

The "calm heating" of zero gravity also works in reverse—there's "calm cooling" as well. When cooling a molten metal into solid form, the solid part obviously will be colder than the liquid section. On Earth, this temperature drop stirs up convection currents. The turbulence disrupts atoms as they try to link up into a solid pattern. The atoms can't line up in the perfect lattice-like grid predicted by theory, because heat-driven currents are knocking them around and creating gaps in the grid.

An alloy with an absolutely perfect grid of atomic bonds might be tremendously stronger than alloys created on the ground. And if all the atoms in that perfect grid can be aligned with their magnetic fields pointing in the same direction, the alloys could have stupendous magnetic properties.

Beyond universal mixing and calm heating, there's a third property of zero gravity that is important to metals research: Samples can be processed without containers. Containers are a source of contamination. Atoms from the container walls will seep into molten metals and change their chemical makeup. And at very high temperatures, the container walls simply start to melt.

In orbit, metals can be levitated and kept in place with magnetism. Or they can be put into an inert gas and levitated acoustically with sound waves. Either method allows containerless processing with zero contamination and no upper limit to temperatures.

Because of contamination, scientists have never been able to test absolutely pure versions of many alloys for their properties. And because containers tend to melt, scientists have virtually no data for liquid metal properties at temperatures above 1,000 degrees centigrade.

Container walls also interfere with growing metal crystals. A perfect crystal must start from a single point, but container walls provide many such "nucleation sites." Competing crystals start to form, making a large, unified metal crystal impossible. Metals grown as a single crystal may be much stronger, more magnetic, or better conductors than ordinary alloys.

At the other extreme are metals with no crystals at all. A completely random arrangement of atoms results in a glass, and scientists are excited by the prospect of metallic glasses from space. The key is to heat metals to extremely high temperatures—which is possible in containerless processing—and then rapidly solidify them before the atoms can line up in any kind of structure.

Glassy metals often are much stronger than their normal versions. A glassy version of aluminum created at the University of Virginia in 1988 tested as three times stronger than steel and with twice the tensile strength of regular aluminum.

PROFITS FROM SPACE FACTORIES

Huge orbiting factories stamping out glassy metals are unlikely, because shipping raw materials would cost too much. But small factories for other products are quite possible. The key is making products that are so valuable per pound that they justify the expensive transportation.

High-value space products likely will include pharmaceuticals, electronic crystals, and high-purity glass for laser lenses and fiber optic cables.

The most lucrative space products are drugs and biological materials. A pound of interferon (to treat cancer and viral infections) can be made into about 4.5 million individual doses. At $10 per dose, a pound of interferon is worth $45 million.

The pioneer in space pharmaceuticals is McDonnell Douglas Astronautics in St. Louis. It's found an efficient way to purify biological materials in orbit. McDonnell Douglas believed this technique could produce a drug, erythropoeitin, to treat anemia, which is a shortage of red blood cells that carry oxygen from the lungs to every cell in the body. Anemia therefore can weaken the entire body and afflicts about 1 percent of the U.S. population.

Erythropoeitin normally is produced by the kidneys to control production of red blood cells. People with diseased kidneys don't produce enough of the hormone to prevent anemia. They often must receive blood transfusions to survive. If erythropoeitin were available to boost red blood cell production, doctors could perform fewer transfusions. A patient would get new red blood cells from his own bone marrow instead of from someone else's blood. Patients wouldn't risk death from blood-carried diseases like acquired immune deficiency syndrome (AIDS) and hepatitis.

McDonnell Douglas spent $21 million to develop its purification process, known as continuous flow electrophoresis. Technicians start out with a beaker or bucket full of the liquid to be purified. This "soup" holds the target hormone, enzyme, or cells along with unwanted contaminants and water.

The soup is fed into the bottom of a liquid-filled chamber. As the soup rises to the top, it passes through an electric field that separates the various parts of the soup into different streams. The flow fans out according to the electric charge of each part of the mixture. A series of exit holes at the top captures the separated streams. The electrophoresis process concentrates the target chemical or cells in one exit hole and sends the impurities to exit holes on either side.

The electric current that separates the mixture also tends to heat the fluid carrying it from the bottom to the top. In Earth gravity, the heating causes convection currents. These churn up the smooth flow and prevent the even separation of elements into orderly streams. On Earth, the power level on the electric field must be kept very low to avoid heating and convection.

In space, heating doesn't create convection currents. Density differences caused by heating are irrelevant when there's no gravity to cause lighter parts to rise and heavier parts to sink. Freedom from convection lets McDonnell Douglas turn the power up high for faster, more accurate purification.

The McDonnell Douglas effort had an upbeat start. The company was the first to negotiate free shuttle flights from NASA under a joint endeavor agreement. The agreement gave NASA the right to test its own compounds in the company's apparatus. And McDonnell Douglas signed up a partner in the drug industry, the Ortho division of Johnson & Johnson. Ortho put in $8 million of its own cash to develop the hormone. McDonnell Douglas and NASA trumpeted the good results from early flights: Space purification of erythropoeitin was 800 times faster than the ground rate and produced four times the purity.

The project has since collapsed.

Academic experts say McDonnell Douglas didn't do its homework on how to use electrophoresis on the ground. They claim the company couldn't have designed a worse piece of equipment to do ground-based electrophoresis. The badly designed equipment made the space operations look better than they actually were. For example, the mixture to be purified enters at the bottom and must rise to the top for capture; the experts say simply turning the apparatus upside down would improve its ground efficiency. They say the poor design inflated the gains claimed for zero-gravity processing.

Another blow came in August 1985 when Ortho Pharmaceuticals quit. Ortho dropped out of the electrophoresis project because it could bring the hormone to market faster with a different ground-based process. Ortho has licensed a genetic engineering process for making erythropoeitin from the Amgen Company of Thousand Oaks, California.

The partnership with Ortho failed from the primary threat to all space manufacturing plans: ground-based competition that moves more quickly than space-based operations. Extremely long waits between flights often cripple space research—promising initial results are put on the shelf for months or years until follow-up experiments can be flown. Ortho decided that depending on the shuttle was too risky even before the *Challenger* tragedy, because space manufacturing payloads got lowest priority when launch delays forced the reshuffling of payloads.

McDonnell Douglas appeared to have salvaged the project in November 1985 when it found another drug company, 3M's Riker Laboratories. But after initial enthusiasm for the erythropoeitin project, Riker reached the same conclusion as Ortho. Riker decided ground-based genetic engineering would get the hormone to market faster than the space project.

The *Challenger* disaster struck another blow. The prototype drug factory scheduled to fly in July 1986 was postponed four years—almost three years of delay from the shuttle grounding and a year or two extra from the crush of military payloads claiming first place in line.

McDonnell Douglas pushed ahead, at one point employing more than 100 scientists and technicians. But by 1988, the company threw in the towel and donated its space processing hardware to NASA. The equipment includes small units for processing material in the shuttle's middeck crew space, and larger "factory" units that McDonnell Douglas had hoped to fly in the cargo bay.

Space-manufactured erythropoeitin may still find success because it has one key advantage: greater purity. By screening out unwanted by-products most effectively, space-made drugs will have fewer side effects than those created on Earth.

Sales of a ground-based drug will plunge if a safer space-purified version is available. Safety is so directly related to purity that ultra-pure drugs likely will drive out second-class versions.

Erythropoeitin may be manufactured in a two-step process: Genetic engineering produces a tank of the material on the ground. Then it is ferried to space for high-speed, highly effective purification.

The market for this drug is huge, because Ortho Pharmaceuticals wants to push it beyond kidney-related diseases. Ortho reported in October 1988 that erythropoeitin eases the side effects of the anti-AIDS drug AZT. Patients on AZT who also received erythropoeitin had a 60 to 70 percent reduction in the number of blood transfusions they needed to combat anemia.

Enthusiasm for electrophoresis is still strong in Europe. The French space agency CNES paid $250,000 to the Soviet Union for the use of a Photon capsule in early 1989. The flight is the first step in a six-year, $31 million program to develop electrophoresis in space. The process can purify almost any biological product, not just erythropoeitin.

CNES plans to install an electrophoresis unit in the Soviet *Mir* space station in 1991 and in the U.S. space shuttle in 1993. CNES plans to have space-qualified equipment ready for the European module of the U.S. space station in the late 1990s. Backers of the project include the French aerospace firm Matra Espace and the French drug company Roussel Uclaf. Spanish and Belgian firms also are investing in the deal.

SUPERSPEED ELECTRONIC CHIPS

After drugs, the next most-valuable space product will be new semiconductor chips for faster electronic circuits. The new superspeed chips will use crystals of gallium arsenide. Some gallium arsenide crystals are grown on the ground now, but they have many defects caused by gravity-driven convection currents. Space factories could grow gallium arsenide crystals to commercially useful large sizes completely free of defects.

Pure gallium arsenide from space may be worth $500,000 per pound. (Today's defect-riddled gallium arsenide sells for about $100,000 per pound.) This high value makes gallium arsenide the only nondrug substance already slated for on-orbit production.

A tiny four-man firm called Microgravity Research Associates (MRA) is leading the charge on gallium arsenide. The payoff with gallium arsenide is a circuit that operates with blinding speed. Chips that replace today's standard silicon with gallium arsenide could work more than 100 times faster.

MRA's executive vice president is Russell Ramsland, Jr., who recalls that the company got its start in a pub while he was attending Harvard Business School in 1978. He was listening to his friend Brian Hughes explain how commercial opportunities could be exploited in space. Also listening was a *Washington Post* reporter doing a theme story comparing Harvard M.B.A.s with Stanford M.B.A.s.

"When her article appeared in the *Post*, it said that at Stanford the M.B.A.s were all agog about Silicon Valley but at Harvard, Brian Hughes was all wound up about space commercialization," Ramsland says. NASA took notice, and asked Hughes to organize a group of Harvard students to research space commerce.

"We got a consulting study going with NASA for the entire second year of business school," Ramsland says. "They gave us some money, and we were supposed to go out into corporate America and find out how to make space commercialization take off. It was a lot better than going to class."

The study reached four conclusions. The first was that most of big corporate America didn't want the risk of high-tech space research. "Their philosophy is, gosh, let somebody else do it and if we like it, we'll buy it."

The second conclusion was that a three-way partnership among industry, academia, and government could be very productive. But industry wasn't talking with the universities, and both of them worked with the government only as contractors, not as partners.

The third conclusion was that NASA's joint endeavor agreement (JEA) could be the tool to create a good industry-academia-government triangle. The JEA trades free flights on the shuttle for a share of the research results. "What NASA really ought to do is find good entrepreneurs, introduce them to their best scientists, and get out of the room. Let the entrepreneur and the scientist decide if there was a business deal there. . . . And if there was, then they could come into NASA and ask for a JEA."

The fourth conclusion was that there were not too many near-term things to commercialize.

During the course of the study, Ramsland met Dick Randolph, a retired Air Force colonel in Coral Gables, Florida. The colonel had bought several cheap Getaway Special cans to be flown on the shuttle, and NASA wanted to know why. Ramsland discovered that he and Randolph had independently come to the same conclusion about space commerce: Crystal production would be a hot area.

NASA then directed them to Harry Gatos at Massachusetts Institute of Technology (MIT), probably the world's leading expert on space crystal growth by virtue of his 1973 experiments aboard the temporary *Skylab* space station. The trio worked out a contract and then sought a JEA.

"We started negotiating with NASA in the fall of 1979, and in a quick three and a half years, we finally ended up with a joint endeavor," Ramsland recalls. The MRA founders then expected to wait at least seven years before the first income started to flow from space production. But they also believed that solid ground-based research on crystal production would pay off even before the first space flights. And it has—MRA has patented an advanced process for making gallium arsenide and other crystals in full gravity.

The other key is that both Randolph and Ramsland had other income. They didn't depend on MRA for salaries. Ramsland helped run his father's oil-drilling business in Midland, Texas, and MRA was the type of long-shot gamble that oil men were willing to make.

MRA's first objective was to put some money together, set up a relationship with MIT, and secure a JEA with NASA. The trio raised $100,000 to pay legal bills and travel expenses, and completed the initial phase in four years.

Intensive ground-based research was phase two. MRA needed $1.5 million to pay for this three-year stretch, and aerospace experts like Wolfgang Demisch, then at the investment bank of First Boston, declared the company had no chance of raising that much money. MRA succeeded anyway. Ramsland notes that ground research is essential to space projects.

"We realized that space was too complex and too difficult to go up and just do research. We had to accomplish everything we possibly could on the ground first. Then we can go up with a clear set of goals. We still have a lot of ongoing research for the next steps, but we've solved all the big technical problems."

The MIT connection was crucial to the company's technical success. "We wanted, from the beginning, NASA and MIT's and our interest all to be the same. And that is, dependent on success, not guaranteed payments. So rather than pay MIT a big upfront chunk for a royalty on the exclusive use of a patent, we agreed to give them a royalty on the gross proceeds less flight costs," Ramsland says. The formula makes the royalty pretty much dependent on profits.

The MIT connection gave MRA cheap graduate student labor, free experimental equipment left over from previous projects, and the broad industry contacts that Prof. Gatos has from his consulting work with other companies.

"We found there were two kinds of researchers," Ramsland says. One kind would grow every type of crystal except gallium arsenide, in order to build a perfect theory of crystal growth. Only after refining the theory would the researcher test it on gallium arsenide.

"But what Harry [Gatos] did, and what we really liked, is ask why [electronic] devices fail—what is needed in the devices market? What are the *relevant* problems in materials that bother devices. . . . Let's just cure what's important. And then let's go to work on them. It may be seven or ten years until you go to market, but at least you are working on problems that go toward a market, like Bell Labs did."

MRA is just beginning its third phase, creating space-worthy equipment to make sample quantities in orbit. The company needs to raise about $18 million to cover third-phase expenses.

The first crystal factory will be about the size of an office copier. The later full-scale production system will be a box almost 13 feet long. The crystal factories will be mounted on an all-purpose NASA carrier in the shuttle orbiter's payload bay. Cost to fly the 7,500-pound factory may amount to $11.25 million per flight. (That's at the shuttle's subsidized price, not its actual higher costs.) With 50 pounds of finished crystal wafers returned each flight, the transportation cost alone — to haul the factory back and forth from space — will be $225,000 per pound. When MRA can permanently leave the crystal factory in orbit on a space station or unmanned platform, Ramsland expects costs to drop by a factor of three.

The space phase of MRA's effort is designed for a slow start. Small sample quantities of space-pure gallium arsenide, and other related crystals, will be sold to circuit designers. Then the designers may take months or years to create new chips based on large-scale gallium arsenide. Only then can MRA gear up for big production runs.

The potential markets include high-power lasers for communications or defense. Today's technology can create defect-free laser crystals, but they are limited to very thin slices. The crystals are grown on a base, or substrate, that is full of defects. When laser power pumps through the crystal, the defects in the base start to move — into the formerly perfect crystal.

"The more power you put through," Ramsland says, "the quicker this junk moves around and destroys your laser, to the point that you can measure a laser lifetime, instead of in years, in picoseconds."

MRA's gallium arsenide would be ideal for undersea fiber optic cables. These cables will need high-power lasers for repeater stations, lasers that can work for years before failing. Laser crystals grown on space-pure substrates would have the required long life because no defects would seep into the crystal from the base.

Another application for pure gallium arsenide is HEMT (high electron mobility transistor) technology. HEMT circuits push materials to their theoretical limits on electron speed, and that's where gallium arsenide shines.

"To give you an idea of the speed differences we're talking about, silicon can move its electrons about 12,000 square centimeters per second per volt. Gallium arsenide with HEMT technology will go a million and a half, not 12,000. We'll get totally new leaps of technological capability," Ramsland says.

Ground-made gallium arsenide is too crude for tiny HEMT devices. In large devices, the random defects in current gallium arsenide even out. Each device will have about the same number of defects, and thus all the devices will behave the same way. But when sliced into the tiny pieces needed for integrated circuits, the defects don't average out. Some gallium arsenide parts will have a lot of defects and others only a few. They're no longer uniform, and quality control becomes impossible. Space-pure gallium arsenide could create the uniform quality needed in tiny circuits.

Battling M.B.A.s and Their Success Formulas

"Most U.S. companies now are being run by M.B.A.s instead of scientists, and it's been a real detriment," laments Ramsland, a Harvard M.B.A.

"M.B.A.s are not geared up to understanding what it takes to win in technology. They are not going to invest in a long-term project, unless you can show them a market study for how many 'widgets' they are going to sell in three years. I walk in and say, 'Hey, I'm going to make a great gallium arsenide' and they say, 'Oh, well, what's your market demand right now?' Well, there isn't any yet. Case is over, pal; you can forget it because they don't believe that good long-term R&D [research and development] pays off."

"What the Japanese have proven, and what Bell Labs proved, is that it does. But these M.B.A.s haven't seen that, they haven't experienced that, they weren't taught that in business school. They've been taught to sit there and look at a market study and to figure out the internal rate of return. If it beats their hurdle rate, invest. If it doesn't, don't. The strategic position aspect is neglected."

"We can describe the market, and we can show you how the economics are going to work. We can tell you, look, we have faith that it's going to be there because the history has always been [in electronics] that you want more power, more reliability, more speed, more temperature tolerance, more radiation resistance. . . . and that's what we offer. But if we wait until the demand actually exists, we're going to be another three or four years developing the space gallium arsenide. We've got to stay on track now to have the product hit somewhere in the time frame that the demand does. And that just does not sell well [among MBAs]."

The company's devotion to thorough ground research has paid off, as Ramsland predicted. MRA expects to complete an Earthside factory for gallium arsenide in one-inch diameters. The one-inch size is only useful in laboratory and research applications. Commercial applications will require a 3.5-inch wafer, the standard size for existing silicon crystals. But the research market still may amount to $10 million a year in sales. The ground-based factory will produce other exotic semiconductors, perhaps including gallium aluminum arsenide.

MRA still expects to go into space for production of the large-scale gallium arsenide crystals required for commercial applications.

Moving into production will require more cash, but Ramsland doubts MRA will raise it through public stock sales. The reason is the pressure for immediate earnings faced by publicly traded firms. Ownership more likely will be spread to a small core of sophisticated individual investors and perhaps foreign firms such as C. Itoh of Japan that understand strategic R&D.

"Increasingly, space represents for me the kind of opportunity that the oil business did for my father back in the 1940s. So we are transitioning a little out of one and into the other."

More Firms to Build Space Factories

The drug factory donated to NASA by McDonnell Douglas and the crystal factory from MRA likely will be the first space manufacturing units. In addition, two more companies have won JEAs with NASA to make crystals: Grumman Aerospace Corporation and International Space Corporation. Both plan to fly crystal furnaces that operate at much higher temperatures than the MRA unit. That may be a drawback, requiring more electrical power and more elaborate ways to get rid of the waste heat.

The high-temperature furnaces may be aimed at new crystal materials with important military applications. The Pentagon plans to fly its own Spacelab mission in low-gravity science in 1990—before any civilian Spacelabs can return to the shuttle—and crystal units probably will be aboard. The military is especially keen to grow a mercury-cadmium-telluride crystal. Such a crystal would boost the sensitivity of space-based sensors for the Strategic Defense Initiative.

NUCLEAR FUEL FROM SPACE

Research into glassmaking could lead to orbital factories making fuel for nuclear fusion. The fuel would be tiny hydrogen-filled glass balls for a type of fusion called inertial confinement. The process resembles blowing up a series of extremely small hydrogen bombs.

In actual hydrogen bombs, a sphere of fission bombs creates intense heat and pressure around a core of hydrogen, heating and compressing it into fusion. Fusion power plants can mimic this with a bank of powerful lasers that blast a tiny glass sphere from all sides. The explosive vaporization of the glass shell heats and compresses the hydrogen inside to fusion levels.

Making the glass spheres in space allows them to be perfectly round. The more perfect the sphere, the more evenly and completely the hydrogen is compressed. The power from each nuclear fusion is brief, but it can be repeated infinitely. This type of fusion resembles an automobile, where tiny amounts of fuel are injected into an engine cylinder and exploded over and over again.

The fusion heat could drive a turbine to generate electricity. A fusion power plant would not have the dangers of today's fission units. The hydrogen fuel isn't radioactive, unlike the uranium that must be mined and enriched for fission plants. The process can't run out of control because any disruption would cut off the precisely timed arrival of the glass pellets. The radioactive leftovers are less toxic—they lose their radioactivity about 50 to 100 times faster than wastes from fission plants. And the fusion wastes are metallic, which means that contamination from a fusion plant won't enter the food chain, through milk from cows, for example.

The hot expanding gases from a fusion chamber also make an ideal rocket engine. Laser power could be beamed from the ground or orbital stations, as Chapter 4 describes. But instead of simply heating a fuel, the lasers would fuse it for vastly greater thrust. The rocket exhaust would be radioactive, however, so fusion engines wouldn't be used in the Earth's atmosphere. Their task would be interplanetary transport, or even interstellar voyages.

Laser fusion is about to reach the stage where experimental rigs produce as much energy as they consume—the energy break-even point. Much more work is required to reach the cost break-even point, so orbital factories for fusion spheres are a bit in the future.

Steve Dean of Fusion Power Associates estimates mass production of spheres won't be needed for perhaps 30 years. On the other hand, perfectly spherical shells or glasses with space-unique properties could provide a shortcut to commercial fusion power. Two professors at the University of Missouri-Rolla, C. S. Ray and D. E. Day, are looking for that shortcut. They melted prototype glasses on the German Spacelab mission in November 1986, and proved that microgravity is an easier place to create glass spheres. But they also had problems keeping a bubble in the center of the sphere, so they need to try again.

The Energy Department is spending about $150 million a year for the overall laser fusion program. Most work is carried out by the nation's weapons labs because the physics of imploding glass pellets with lasers is very similar to using fission bombs to implode a hydrogen bomb.

SPACE MANUFACTURING: SCIENCE IN SLOW MOTION

University and corporate pioneers alike have long suffered from a frustrating inability to fly experiments as often and as quickly as they need to keep ahead of ground-based competitors. During the Apollo era, materials processing got top billing only during the 1973 *Skylab* flights. Then for ten years a Dark Ages of space science reigned, until the shuttle started flying in 1983 with crews large enough to conduct experiments.

The shuttle space manufacturing Renaissance wasn't ideal—materials processing experiments were immediately dumped from the shuttle schedule any time NASA could snare a communications satellite customer. Even so, the Renaissance was only three years old when the *Challenger* disaster struck. Another Dark Ages descended. And when the shuttle flew again, high-priority Defense Department payloads crowded out most space manufacturing work for the rest of the 1980s.

In pre-*Challenger* flights, room on the shuttle was very scarce for materials experiments. The prime area was the crew section, in an area called the middeck. Researchers with packages ready to fly were kept waiting for a year or more—they were continually bumped to later flights to make way for communications satellite customers.

For example, NASA and Arianespace fought fiercely to win the launch contract for Arabsat, a communications satellite owned by 21 Arab countries. NASA defeated Arianespace by agreeing to take Saudi Arabian prince Sultan Salman Al-Saud along to view the deployment. Carrying this extra crew member required stuffing more food and equipment in the middeck, and experiments scheduled for that flight were tossed off. NASA's success with Arabsat also meant taxpayers were forced to subsidize Muammar Qaddafi's Libya and the Palestine Liberation Organization, two of Arabsat's owners. Arabsat paid NASA just $11.6 million, while its share of the cargo bay cost the government up to $90 million.

In another case, middeck experiments planned for an August 1985 flight were scrubbed because NASA mounted an emergency salvage mission for another of its comsat customers, Hughes Aircraft. The equipment needed by astronauts to retrieve Hughes' errant satellite took up the middeck lockers.

A post-*Challenger* policy forced most commercial comsats off the shuttle to expendable boosters. Experimenters ought to get more respect now. On the other hand, the Pentagon has made noises about claiming a large share of the middeck space for its own experiments.

Thus, during the 26-year manned space program, American researchers have had real access to space for only 4 years—the brief 1973 *Skylab* flights and the 3 years of initial shuttle flights.

This means that materials processing in space is not well researched. It's been such a career wasteland that only a few American academics and businessmen have been working the area. Many more corporate and academic players are needed to simply map the rewards of space manufacturing, and still more people will be needed to actually harvest them.

The only way to attract these new players is with routine and continuous access to space. NASA's attitude toward materials research is still rather aloof, judging from the post-*Challenger* payload schedule. NASA is giving prime early launch slots to NAVSTAR satellites that could just as easily be launched by expendable rockets. At the same time, Spacelab missions — which can *only* be flown on the shuttle — are postponed for years.

Even if the NAVSTAR satellites were dropped from the shuttle manifest, materials experiments would still be fighting for space with Defense Department and planetary exploration probes. NASA has no plans to ease this bottleneck with expendable rockets. Perhaps the most positive sign for materials science was the scheduling of several experiments for the first post-*Challenger* flight. Five drug companies flew experiments in protein crystal growth (which is discussed in Chapter 8).

Until NASA schedules frequent flights for materials research on the shuttle or expendable boosters, we'll be left guessing about its true value. There simply won't be enough investigators willing to cripple their careers doing slow-motion science to probe all the possibilities.

3M: Looking for Profits in Zero-Gravity

A meeting in St. Paul, Minnesota, a few weeks before Christmas 1983 may have marked the beginning of the post-NASA space age. The top brass at 3M had gathered to plot how their new long-range research center would keep them competitive through the end of the century.

3M, at that point, had $7 billion in worldwide sales. The company made everything from sandpaper to Scotch tape to specialty chemicals. It wasn't an aerospace contractor; the closest 3M link to the shuttle program was NASA's attempt to protect the orbiters' tiles by spraying them with Scotchguard.

Chris Podsiadly, the research lab's director, explained that his scientists had found problems in some new technologies they were exploring. They couldn't seem to scale them up to anything approaching useful size.

One answer would be to run some experiments aboard the space shuttle, because the crystals and thin films they needed might grow larger away from the disruptive influence of gravity. Dr. Podsiadly looked around to see how the assembled top executives would react. The chairman of the board, the president, the vice president for R&D, and the vice president for technology were being asked to do something quite unusual.

"What do you want us to do with this idea?" Dr. Podsiadly remembers asking. The executives said to start work on it.

"I asked them, 'How *far* do you want us to go with this?' And I remember [chairman] Lew Lehr saying, 'All the way. This looks really good.'"

3M became the first major corporation—other than aerospace contractors living off NASA contracts—to push space-based R&D. The corporation found it *needed* the advantage of low-gravity research to protect its technological edge from competitors. Executives realized that space provided a new research tool as basic as a Bunsen burner. Just as the Bunsen burner tests the reactions of chemicals to the addition of heat, the shuttle would let 3M test what happened with the subtraction of gravity.

Dr. Podsiadly formed a corporate commando team with volunteers from various parts of the company, because the project lacked any formal budget. The team's main challenge was to design and build equipment that would turn science problems into formal experiments. NASA officials encouraged them to look forward to their first flight on the shuttle... in perhaps three or four years.

The commando team aimed to reach space much sooner.

"Our whole idea then was to fly in a year. NASA told us we were crazy," Dr. Podsiadly says.

Two months after top executives had blessed the space plan, 3M signed its first pact with NASA. It was a "memorandum of understanding," the bureaucratic equivalent of going steady. Nothing is binding on either party. But the agreement signaled to NASA employees that 3M should be taken seriously.

By April 1984, the 3M strike team numbered in the dozens, many working 14-hour days, seven days a week. They produced their first engineering drawings of the initial experiment, but attempts to get on a shuttle flight were hitting dead ends.

By summer 1984, 3M researchers were building the experiment. They visited NASA centers for help and kept getting the same question from officials: "What's your JEA number, so I can charge back this work?" Clearly, a JEA (Joint Endeavor Agreement) was something the bureaucracy expected 3M to have.

JEAs are joint ventures between NASA and a company. NASA provides free shuttle flights and expertise in return for access to the information a firm develops. NASA also hopes JEAs will lead to new customers who can pay full price to use the shuttle. The 3M lawyers went into overdrive to get a JEA, aided by NASA's Mike Smith, who provided essential help at the headquarters level. In the meantime, the 3M technical team kept telling NASA people, "I don't have a JEA number. Gosh, do I need one?"

They delivered their hardware to NASA's Marshall Space Flight Center on September 5, 1984, for shake testing and structural analysis. A few days later, the lawyers came up with the JEA. The experiment flew in November. The nine months from first NASA agreement to flight was unprecedented — a spectacular feat compared to the three years required by experienced aerospace contractors.

The speed is a crucial sign to other potential researchers that the agonizingly slow pace of space science can be overcome. Researchers now know they don't have to be satisfied with only a few flights per decade. 3M showed how space science can be forced into a much faster pace.

3M is reluctant to speculate on what actual products will come from space-based research. The company is doing very basic chemistry, executives explain, and is far away from practical applications. But they also hope for great discoveries and aren't willing to tip their hand to competitors. For example, they recently persuaded NASA to extend the secrecy period on their experiments to two years, up from one year previously.

Any number of strange new products might spring from the firm's space work. 3M could develop a "smart" adhesive, one that would be sticky only on command. "Smart" adhesives and chemicals could result from 3M's investigation of how molecules decide to order themselves. Once that is known, molecules might be commanded to shift position to present their "sticky" end to another surface. The process might work in reverse, allowing things to become unglued on command.

Another possibility is super-strength plastic that keeps its hardness even at high temperatures. Automakers are keenly interested in this work. General Motors plans joint space experiments with 3M to learn how to create plastics that reach more of their theoretical potential.

Optical computers and fiber optic communications are the most lucrative applications for 3M's work, however. Present computers are driven by electrons moving through silicon. Upcoming computers will replace silicon with gallium arsenide because it allows electrons to move at higher speeds (see Chapter 7). But after gallium arsenide, computers will go optical with laser-pulse information moving at the speed of light. Experts hope for devices with 10 to 100 times the speed of electronic circuits. The challenge for this optical generation is developing the switches and amplifiers needed to manipulate light-based data flows.

Organic crystals—the target of 3M's first experiment in space—can combine two incoming beams of light, for example, and put out a beam that is the sum of the two incoming frequencies. Being able to "add" is the fundamental basis of most computers.

Optical computers, pushed by the raw computing demands of the Strategic Defense Initiative, are expected in the early 1990s. A more immediate market for 3M's optical work is fiber optic communications. Fiber optic cables are seriously challenging communications satellites as the low-cost champion for voice and data transmission. What's lacking is a good method of optical switching and amplification for these systems. 3M's organic crystals could serve a massive demand in just a few years.

The quality of the crystal in a switch or computer circuit determines how much of the incoming light is smoothly converted to the output beam. Defects will turn part of the incoming beam into waste heat, threatening any optical computer with meltdown.

3M researcher Bill Egbert was wrestling with organic crystal growth in late 1983. Egbert could produce organic crystals in powder form and had encouraging results. But he couldn't grow large-scale crystals to confirm his experiments at a practical working size.

Another 3M researcher, Mark Debe, was vexed with a similar problem in creating thin organic films. He could deposit an initial layer of molecules on a base material, and the molecules seemed to line up to match the molecules in the base. Getting organic molecules to "self-order" like this could create properties that ordinary randomly arranged molecules couldn't offer. The problem was that depositing additional layers didn't work. The extra layers didn't repeat the order created by the base material.

Then the accounting firm of Coopers and Lybrand made presentations to 3M executives on the benefits of space work. (NASA had hired Coopers to try to drum up customers for shuttle and space station microgravity facilities.) When Debe and Egbert considered the Coopers and Lybrand pitch, they recognized that a low-gravity environment could eliminate the turbulence that was ruining their attempts to make larger crystals and thin films. Growing organic crystals and films could be done more gently in space, allowing weak forces that are overwhelmed by gravity to take over and create self-ordered materials.

The researchers' problems led to the pre-Christmas decision by 3M's top management to enter space R&D.

"The ideas we had were extensions of our long-range ground based research," Dr. Podsiadly says. "The main driving force was that we could probably achieve our objectives faster, and make our decisions better, because the information coming in would be something that we couldn't do here."

"The economic part of it is easy to explain," Dr. Podsiadly says. Eliminating gravity often can simplify experiments so that researchers can concentrate just on the effects they want to study. "That might cut a year off long-range ground based research. That's big money. Research in today's environment is extremely expensive."

Dr. Podsiadly and others at 3M refuse to say how much their space lab is costing, or to reveal the general expense for a typical space experiment. (Outsiders estimate the space lab takes several million dollars a year to run.) They aim to become an independent business unit within 3M—perhaps selling their space research expertise to other companies, perhaps running a space station lab for profit, perhaps actually manufacturing in space. They aren't about to give away the crucial cost data they've sweated to accumulate through countless seven-day weeks.

The success of their blitzkrieg approach to space science makes them the unquestioned experts in getting the best results from modest investments.

BUILDING THE FIRST MACHINE

The team first tackled organic crystal growth. Their lab bench in space was a canister 19 inches in diameter and 2 feet tall. It was triple sealed — the "can-inside-a-can, inside another can," assured NASA that the solvents 3M wanted to use would stay put. Earl Cook, a physicist and the team's hardware and electronics wizard, explains some of the work obstacles:

"The biggest thing that we found was not the difficulty of the task but knowing what had to be done," Dr. Cook says. "You'll find some NASA people who say one thing about how you build hardware, and then you'll ask somebody else and they'll say something completely different. We found that the hardest part."

"One of the big barriers we came up against was learning what NASA documentation was all about," Dr. Cook says. "NASA is very, very closemouthed on what they expect in terms of documentation."

The 3M team had a shock when it delivered its first experiment to Kennedy Space Center (KSC). KSC's first action was to assign them a quality-assurance person. And the first thing he said was, "Let me see your shipper."

"What do you mean, 'shipper?'" Dr. Cook remembers asking.

A *shipper* is a document that traces a piece of hardware back to its birth from separate flight-certified components. It proves that the assembled piece is flight-certified hardware.

"Nobody had ever told us that we'd need a shipper. When I told the [quality-assurance] guy I didn't know what a shipper was, he just said the hell with us and locked the door. That's the last we ever saw of him. We got a new guy the next day."

The next time 3M arrived with an experiment, the shipper met NASA's approval. "We *nailed* it," Dr. Cook says fiercely.

The exacting certification process only applies to parts outside the triple-walled container. The outside parts are the ones that could hurt the crew or damage the orbiter.

"Those components on the outside have to meet certain MilSpecs [military specifications]. We have to say where we got them from; we've got to say what we've done to them. Those things on the inside, they could care less."

Dr. Cook fought to get a battery included in the first experiment so he could power a real-time clock, and lost. The safety committee's battery expert from Johnson Space Center, Bobby Bragg, said batteries had to have special containers, with special coatings in case they ruptured, and tubes to vent any escaping gases overboard. It was a massive engineering task that would have hopelessly complicated 3M's goal of quick, straightforward experiments.

For the second flight, Dr. Cook tried again. He showed that the equipment had a fuse to shut things down if the battery ever shorted, and a diode to make sure the current never reversed and started charging the battery. (Accidental charging could cause the battery to explode.)

This time, Dr. Cook won the battery argument, but only because Bragg had retired in the meantime. "He had over the years become the expert in batteries and nobody ever questioned him," Dr. Cook says. NASA employees are unwilling to challenge each other because the agency has become very compartmentalized, Dr. Cook found. That's one reason 3M received so many conflicting answers to its questions about requirements.

"NASA is a traditional bureaucracy. You had trouble finding people willing to make an independent decision. They've always got to go one level up. That's the way bureaucracy works—no one is willing to take a chance or a risk. I think in industry you'll find more people willing to make an independent decision," Dr. Cook says.

3M was aided immensely in meeting its schedules by allies within NASA who stepped in with crucial help when paperwork threatened to abort an experiment. The 3M team is grateful for these allies but wishes that guiding angels weren't necessary so often.

The ultimate paperwork horror struck after their third flight. By then, 3M had learned that one thick stack of documents—the Flight Operations Support Annex (FOSA)—had only a single use. NASA officials used the FOSA solely to draw up a chart full of strange symbols and boxes for the astronauts showing hour-by-hour activities they would carry out.

"We did our own chart. We generated our own hieroglyphics and put them in the boxes. We sent it to NASA, and the payload people down there said, 'Yeah, this is fine, we'll accept this'," Dr. Cook says. There was no need for the FOSA because 3M leapfrogged it to produce the actual end product.

But two months *after* the flight—long after the astronauts had any need for schedules—NASA officials demanded that 3M produce the "missing" FOSA. All documents for the flight had be accounted for in the master record, and 3M was short one FOSA. "So we spent two weeks generating the damn thing. That kind of inflexibility is a big irritant."

NASA's astronauts were a different story. Many of them have advanced science degrees and are eager to help with 3M's orbital chemistry set. On the first experiment, though, 3M accepted the bureaucrats' rule that astronauts could give only limited help: They could flip one switch to turn the experiment on. The first 3M experiment flew successfully with just this on-off help.

The 3M team wanted more for their second experiment. They wanted to have nine different chambers turned on in various combinations during the astronauts' sleep periods. With the astronauts asleep, 3M hoped to grow thin films when there would be the least amount of vibration from crew members bounding around. 3M's team wanted to put in a small keyboard and digital display so an astronaut could punch in some codes that might fix something or start a sequence.

"The training people were aghast," Dr. Cook recalls. "They said 'You cannot, *you cannot*, train these astronauts. They don't have time to learn this thing.'"

The 3M team did an end run around the objections, inviting astronaut Ox vanHoften to St. Paul to see the experiment. They explained their proposed keyboard controls and asked whether he'd like to be part of the experiment. Even though he was getting ready for an emergency spacewalk to retrieve a broken satellite, vanHoften was eager to help. Perhaps because of his prominence with the upcoming satellite spectacular, NASA officials went along with his and 3M's request.

The keyboard saved the second experiment, which went on the fritz just before the end of the first night's work. Dr. Cook was in a Houston motel room watching the flight on NASA Select television when vanHoften announced, "We've got an alarm light [on 3M's experiment]." Although the astronaut knew the condition wasn't hazardous, anxious NASA controllers phoned Dr. Cook only a few seconds later to get him to fix it.

The researchers came up with a series of unplanned tests that vanHoften punched in from the keyboard. The computer seemed to be running fine, and no problems showed up with the electronics inside the canister. Fearing cable connection problems, vanHoften even pulled apart all the plugs and reassembled them. Yet when he tried to crank up an experiment, the machine balked. 3M's team eventually guessed that the apparatus was running too hot.

One of the biggest problems in space is how to get rid of excess heat. Hot air doesn't rise—its lower density is irrelevant in zero gravity—so excess heat is hard to drain. The 3M group experiment lost heat only when it slow-ly conducted through the experiment walls into the body of the orbiter. The overheating was solved by running a single chamber in the canister at a time, instead of doing double runs as planned. The keyboard and helpful astronaut saved the experiment from failure.

3M has done two types of experiments in orbit thus far. One is DMOS—diffusive mixing of organic solids. In orbit, an organic solution is allowed to diffuse into an incompatible solvent, causing the organic material to crystallize out of the solution. The lack of gravity allows the crystal to stay suspended and continue growing, rather than falling to the bottom of the container.

But on the first DMOS run, the six units simply did not work as ex-pected. Some grew nice crystals, some grew puny crystals, and some grew none at all. On the DMOS-II experiment in November 1985, 3M scientists were determined to find out why. They rigged two chambers to simply hold dyes—no crystal-growing solutions. The movement of the dyes from one compartment in a chamber to the next would tell them how materials actual-ly diffused on orbit.

The results were clear-cut, according to Chris Chow, manager of the space laboratory. "One chamber had dyes with densities that were almost equal. We opened the valve and let the dyes diffuse. Lo and behold, gee whiz, they didn't completely mix at all. After seven days on orbit, you've got 80 percent left [in their original cells]."

But in the other chamber where the two dyes had a small difference in density, the mixing was complete. The results confirmed that the large crys-tals grew because the starting solutions had densities different enough to cause good mixing.

Density, of course, is a measure of how heavy a material is compared to its volume. (Lead is dense, whereas popcorn isn't.) That's not supposed to make a difference in orbit, but 3M found that it did. Dr. Chow says the "weightlessness" on the shuttle really isn't pure enough to be called microgravity—perhaps it's more of a milligravity. And the milligravity caused enough convection between different density liquids to actually ac-celerate crystal growth. Now 3M scientists wonder how to take advantage of this fact to help other material processing experiments in orbit.

The other experiment—the one vanHoften saved with on-orbit reprogramming—is called PVTOS, for physical vapor transport of organic solids. In one end of each of the nine chambers, an organic solid is heated so that it vaporizes. At the other end of the chambers is a plate on which it condenses. A thin film is created on the plate, and 3M scientists hope it will be self-ordered, with its molecules lined up in a pattern matching the base plate.

3M scientists tried the thin film experiment because they are hungry to learn how molecules decide to attach themselves to a new surface. Better understanding of this would apply to most 3M products, from Scotchbrite floor cleaner to the thin films 3M lays down on its recording tape and photographic film. A better thin coating to make X-ray photo film more sensitive, for example, would protect patients by allowing lower-power, lower-dose X-ray exposures.

A third 3M experiment will try to create plastics with a greater percentage of crystallization. It hasn't flown yet, but it's attracting interest from other corporate giants: General Motors (GM) and Dow Chemical. GM and 3M will work on a study of Nylon 6, a plastic the car maker uses heavily. Earth-bound methods create a Nylon 6 that falls short of its potential—only 45 percent of any Nylon 6 piece is the desired crystalline form.

"There are indications of quite dramatic changes in low Earth orbit," Dr. Podsiadly reports. "That's the prediction now: more crystallinity, more strength."

GM officials don't plan to start building car parts in orbit any time soon. But like 3M, they believe that microgravity research can speed their efforts to understand and thus better control Earth-bound operations.

Dow Chemical is joining the plastics experiment but hasn't disclosed what type of material it will study. With the *Challenger* tragedy, the 3M-GM-Dow plastic tests were greatly delayed.

That's frustrating for Dr. Chow and other space researchers. "We want to fly as soon as possible, and get as much data as possible, so there is feedback, so we can design the next experiment."

3M's space lab avoids complicated all-in-one experiments. It instead creates stripped-down experiments that evolve after flight experience. Dr. Chow says, "We recognize, from a reality standpoint, that upper management is not going to wait for five years and say, 'Have you flown yet? What did you get?' They are not going to tolerate that." So Dr. Chow and the team knew from the start they needed a steady stream of small victories, rather than one bells-and-whistles extravaganza that might not fly for years.

Now the space team has 14 official members and many more volunteers among the other researchers at headquarters. "It's almost like a spy network," says Dr. Chow. Team members reach out and get bootlegged help from associates, and those recruits in turn often seek out still others.

Most academic microgravity work is done by a lone professor with perhaps a couple of graduate students to help. That makes the 3M team more active and well-funded than any university operation.

"Our people have just spent two years getting an advanced education in microgravity research," helping to found a new area of science, Dr. Podsiadly notes.

"Some people argue that you don't create science; you just create technology. Well, bull. We're going to create a new science. There's precedent for that. Polymer chemistry—you can get masters and Ph.D.s in polymer science now—developed when industry in World War II had to come up with synthetic rubber and nylon. The science developed from industry to academics. And I think the same thing is going to happen with microgravity science."

3M WINS TEN-YEAR NASA DEAL

Late in 1986, 3M signed an agreement with NASA to get 62 experiments flown free over the next ten years. The pact gives 3M access to the middeck twice a year. The company gets to fly in the payload bay twice a year in the first three years and six times annually in later years.

3M will allow NASA to use its hardware and laboratories for government-sponsored research. NASA also will have access to much of 3M's information on its private experiments and be allowed to collaborate on them.

Perhaps more important, NASA will get a huge multinational corporation firmly committed to space research. That's important in terms of justifying the cost of the space station. NASA needs to have major potential customers for the space station at work on the space shuttle to show Congress that corporate America needs an orbiting research center. The 3M agreement, with its hooks to GM and Dow Chemical, is exactly the endorsement NASA needs for the space station.

Orbital R&D for
Earth-Side Problems   8

The biggest space employers of the next decade could be orbital R&D labs, working on ways to improve Earth-bound factories and offices. These labs would manufacture information in space, not products. Their fresh insights would be transmitted back to Earth to improve factory efficiency or to create new products and services.

Producing information instead of products beats the problem of high transportation costs—information has almost a zero weight. Once the lab equipment and researchers are on orbit, only tiny amounts of experimental material need be shipped upstairs.

For example, Deere & Company is planning to cast iron and alloys in space. The company wants to learn about the physics of metal formation because much of today's metallurgy is done by cookbook with no detailed understanding of cause and effect. Deere's space research is devoted solely to improving metallurgy on Earth. The firm has no desire to cast John Deere tractor engines in orbit.

GM has a similar goal. After looking over the shoulders of 3M researchers, GM plans shuttle experiments on polymers. GM hopes this will bring improvements in the high-strength plastics it needs for its future cars.

SPEED IS MONEY

Companies pursuing new or improved products can hear two clocks ticking. One clock tolls for the enormous salary costs of an in-house research staff. The other ticks away the useful life of the company's existing products as they gradually become obsolete. These dual time pressures make companies extremely anxious for quick research results.

Orbital R&D labs can speed up research by eliminating many side effects caused by gravity. Scientists can examine the "ideal" experiment, using universal mixing, calm heating and cooling, and containerless processing. By eliminating gravity, turbulence, and container contamination, they can focus just on the basic forces controlling their results. The simpler environment will speed up their work.

Many firms will use orbital R&D labs solely for this extra speed, not because the research is impossible on the ground. For example, small test batches of drugs, alloys, or semiconductors will be produced quickly in space and then tested on Earth. After a "winner" is found, the Earth-side process can be adjusted to come as close to the space version as possible. The quick test batches from space would point the way, and cut months or years of trial-and-error from the development process.

GROWING HUMAN CELLS IN ORBIT

Doctors would love to take human cells and grow them into organs and cancer tumors in test tubes. Potential disease cures then could be tested in the test tubes, instead of in live human beings. But human cells refuse to cooperate — they won't grow into three-dimensional groups or organs outside the body.

The Bioprocessing and Pharmaceutical Research Center (BPRC) in Philadelphia did pioneering work to overcome the puzzle of human cell growth. Currently, scientists can keep some single cells alive while they are suspended in a turbulent bath of nutrients, but the turbulence tears apart any multicell groups. Other types of cells can live if they are anchored to a surface, but that produces only a two-dimensional flat sheet of cells. Scientists need to study cells that grow in three dimensions into complete organs such as kidneys or cancer tumors.

The nine-person staff at the BPRC was sponsored by the University City Science Center, with funds coming from NASA, Pennsylvania's Ben Franklin Partnership, the Greater Philadelphia First Corporation and corporate sponsors. Unfortunately, the BPRC went out of business during the *Challenger* hiatus.

Much of the payoff from this type of research will be information that can be applied to strictly ground-based production. Scientists will be able to run controlled experiments on human cells in space to see how they interact — for example, how cancer or virus activity in one cell triggers changes in nearby cells. The research takes place in space simply because that's the only place where complex groups of human cells may be able to survive outside the body.

In addition to researching cell growth, the BPRC worked on cell purification with a set of equipment that was unique in the world. It had a continuous flow electrophoresis machine on loan from McDonnell Douglas. The machine, one of only two that McDonnell Douglas allowed out of its labs, is the champion at large-volume purification of cells. At the other end of the scale, it has an electrokinetic analyzer that takes a small sample and charts how each part will react to the electric fields in the big purifier. The analyzer, in effect, is the aiming device for the purifier. It predicts what settings on the large machine will isolate the desired cells.

Dr. Jack Sloyer, Jr., the BPRC's assistant director, said the analyzer eliminates an enormous amount of drudgery. The standard way to analyze a "soup" for electrophoresis takes hours: A lab technician peers through a microscope at fuzzy little specks, which are the various types of cells and debris in the liquid. The technician picks one out and times it as an electric current pulls it across the microscope slide. The speed of each speck is clocked one at a time; a technician can analyze ten samples a day with a 10 to 15 percent error rate.

The new analyzer flashes a laser beam across the specks as they are pulled back and forth by alternating electric currents. The movements are fed directly into an IBM personal computer, which quickly plots a graph of all the elements in the mixture, showing their electrical charges. Sloyer could analyze a sample each minute, instead of ten per day, and the error rate would drop to only 2 percent.

Scientists who used the BPRC setup could get almost instant analysis of their mixtures. The soup then could be popped into the BPRC's big electrophoresis machine. Because the quick analysis scan laid out the soup's characteristics, the purification unit could be "aimed" directly at whatever component the researcher wanted to isolate. The BPRC's speed at purification stunned the outside scientists who used the facility.

William Claybaugh, who was space division vice president of the University City Science Center, saw a potential commercial marriage of the BPRC's cell purification and cell growth research. Claybaugh expected that cell growth research may eventually find a human cell line that thrives in microgravity. Then pharmaceutical companies could grow interferon, interleukin 2, and other key biologicals in these human cells — a far safer method than grafting human deoxyribonucleic acid (DNA) onto a mouse cell or nonhuman microbe. Forcing yeast to grow interferon, for example, may not produce interferon that is 100 percent compatible with real human interferon. Spurring human cells to produce the target hormones and enzymes likely will result in more effective drugs with fewer side effects.

Claybaugh expected that technicians on the space station could operate cell growth reactors to produce the target drugs, and then float over to use McDonnell Douglas' electrophoresis machine to purify the batch. If scientists find a cell line that grows vigorously in microgravity, the tremendous efficiency of the McDonnell Douglas machine would allow purification of a year's supply of a biological agent in only a few days.

Another commercial application of the cell growth research is quick on-orbit testing of cancer drugs. These drugs now go through a very drawn-out process that starts with animal testing. Diseased mice, for example, receive various chemicals to see if they kill tumors without killing the mice. Then the drugs are tested in higher animals, perhaps dogs or chimps. If the chemical remains effective and still doesn't have lethal side effects, it may be tested on humans. The chemical may fail to work on human tumors, which inevitably have some differences from animal tumors. And it may turn out to be unsafe in humans, despite safety in animals.

Perhaps worst of all, accurate testing requires that some patients receive useless sugar pills or water injections (placebos) instead of the real drug. These patients may die because they failed to get effective treatment—but without them, doctors would lack a control group to prove the chemical actually works.

On-orbit testing would drastically shorten the entire process by providing actual human cancer tumors. Promising drugs could be tested for effectiveness directly on the human cancer. Experimenters could skip all the work on mice, dogs, and chimps—cutting *years* off the testing process. The drugs also could be tested for safety against normal human cells from the kidneys, liver, heart, and nerves. Creating control groups would be a simple matter of labeling test tubes, rather than withholding treatment from desperately ill human beings.

Direct testing also guards against false information from animal tests. If a drug doesn't work on mice and doesn't work on dogs and chimps, doctors will seldom risk tests on human patients. But some chemicals may work *only* in humans. Animal screening can falsely block their discovery.

Another potential payoff from direct testing is the discovery of effective drug combinations. Cancer researchers are finding that interferon, for example, is sporadically effective when used alone but might be very effective when used in combination with other drugs. Searching out the effective combinations of thousands of drugs is almost impossible if every new compound has to be run through years of animal and human testing. Direct testing, with results in a few weeks instead of a few years, would make broad screening programs affordable.

Claybaugh expected that on-orbit cancer testing may take ten years to bring off—the first cell growth reactor has yet to fly on the shuttle. But if Claybaugh's visions come true, orbital research stations just for cancer testing could employ dozens of doctors, scientists, and lab technicians, in addition to space complexes devoted to other research work. Employment on various R&D space stations could reach substantial levels.

GIANT PROTEIN CRYSTALS

Another medical gain from microgravity is the ability to grow huge (relatively speaking) protein crystals. The giant size of the proteins reveals their geometric structure, to be mapped and analyzed. Medical scientists believe that knowing the structure of proteins will give them clues on how to block harmful ones and encourage or mimic useful ones.

The leading specialist in giant protein crystals is cancer researcher Dr. Charles Bugg of the University of Alabama-Birmingham. He heads the Center for Macromolecular Crystallography, one of the commercial space centers started with NASA grants and continued by corporate contributions.

An example of the center's work is human C-reactive protein, which is part of the immune system. Discovering the physical structure of this protein may give clues on how it works in the body.

Another target is bacterial purine nucleoside phosphorylase (PNP), an enzyme that destroys cancer-fighting cells. Researchers want X-ray diffraction photos of PNP to outline its structure. That may lead to ways of blocking PNP production, and thereby allow anticancer agents to work effectively. Burroughs Wellcome Company, a British pharmaceutical firm with offices in Research Triangle Park, North Carolina, is paying for the PNP research.

Dr. Bugg's center has attracted several other drug companies, including Merck & Company, Upjohn Company, Smith Kline & Beckman, Eli Lily, Kodak, DuPont, Procter & Gamble and Schering Corporation. McDonnell Douglas Astronautics provides major funding to the center.

The center has spawned a start-up firm, BioCryst Ltd., that is trying to turn the center's research into a commercial product for treating AIDS. BioCryst has patents on the product and $5.2 million in funding from local business leaders. The firm is operating from an "incubation center" at the university, a facility for start-up companies that provides them with office space and consulting help from university experts.

The center also attracted an international controversy, when an Australian researcher decided to offer technology developed through the center to the Soviet Union. The researcher, Graham Laver, has been working on an enzyme called neuraminidase that the flu virus needs to invade human cells. If neuraminidase can be neutralized, Laver could cure the common cold by preventing the virus from spreading in a human body.

Laver's research has been supported by the National Institutes of Health, and the Alabama center provided the shuttle flights needed to refine the technique used to grow neuraminidase.

Military officials are aghast at Laver's plan to grow the enzyme on the Soviet space station *Mir*. Germ warfare—using viruses to disable an opponent—is a key part of Soviet military research. An effective way to immunize Soviet soldiers entering an attacked region is essential to Soviet biological warfare strategy, and Laver's U.S. funded research could provide that key. Laver has said that because all his findings will be published openly, they won't give any military power an advantage over the others.

The interest in protein crystal growth centers on research, not production in space. But that still could lead to significant employment in space, with orbital service centers growing proteins for ground-based research and analysis.

MANUFACTURING RESEARCH IN ORBIT

Many factories have gradually fine-tuned their production processes with years of trial and error. Management has fiddled with equipment until it makes products that are good enough—but often without real understanding of the underlying processes.

Gray cast iron is an example of this seat-of-the-pants science. It's an alloy of carbon and iron—flakes of graphite embedded in an iron matrix. The pattern of the graphite flakes determines the alloy's strength and other properties. But the creation of graphite flakes depends on how fast the mixture cools, the impurities it contains, and gravity-spurred convection. Foundry managers can't zero in on the effects of cooling rates and impurities because they can't keep gravity-caused convection under control.

Two European scientists decided that space tests might solve the problem. Dr. H. Sprenger of MAN Neue Technologies and Dr. H. Nieswaag of Delft University of Technology in The Netherlands set up experiments for the German D1 Spacelab mission to find exactly what controls the placement of the graphite flakes. Removing convection through zero-gravity tests would let them plot more clearly and precisely the impact of cooling rate and impurities on the structure of cast iron. With those answers known, figuring out the impact of Earth-side convection would be much simpler.

The German Spacelab flew in November 1985 with 76 experiments, including the one on cast iron. Sprenger and Nieswaag may have to try again, unfortunately; their samples weren't cooled at the correct rates by the onboard furnace. Nevertheless, their work is an example of how Earth-side manufacturers will use space laboratories to improve their factories. Their work also shows how even glamorous space science can be laid low by everyday equipment foul-ups.

The German government recently created a new space institute for metallurgical research called ACCESS. (The English acronym is created from the German words Aachener Centrum für Erstarrungsfrontdynamix unter Schwerelosigkeit.)

Peter Sahm of the Foundry Institute in Aachen runs it. Sahm notes that gravity makes it impossible to accurately study many Earth-side processes that are critical for modern technology. Diffusion is a prime example. The diffusion rate measures how quickly one liquid or gas will seep into another liquid or gas. Diffusion is the mixing that takes place even when there are no currents or stirring. Space experiments have proved that Earth-bound scientists simply can't measure pure diffusion rates. Convective mixing caused by gravity always contaminates the measurements. Space tests show that *pure* diffusion rates are much slower than Earth experiments had indicated.

"Precise diffusion coefficients are very important for solving problems in solidification, in the utilization of chemical reactions involving fluids at high temperatures like salt melts as catalysts, or various surface treatment process of metals, glasses and ceramics," Prof. Sahm says.

Another key Earth process is the shift of liquids into gases. The transition to gas is known as the critical point temperature. Scientists need to measure the precise pressure, temperature, and density of elements as they shift to gas. But gravity skews the measurements because small density differences between the two phases cause layering. In orbit, there's no layering as the critical point is reached. Scientists finally can see the pure form of critical point transitions and get exact measurements. Such measurements will give Earth-bound managers their first precise grip on the precise physics of how liquids boil into gases.

In addition to ACCESS, other German centers for excellence are gearing up. The Space Institute is a collaboration of the Free University of Berlin and the Technical University of Berlin. The Space Institute will develop methods for production engineering on orbit, on the Moon, and on asteroids. The ZARM center at the University of Bremen will conduct studies with the combined faculties of fluid mechanics, material sciences, biotechnology and numerical analysis. The University of Freiburg is specializing in crystals, and the University of Würzburg is setting up for biotechnology and biochemistry.

On the private side, a German-led consortium called INTOSPACE is probably the largest microgravity group in the world. The group was formed in late 1985 after a German consulting firm, Keinbaum Unternehmensberatung of Düsseldorf, said European microgravity needed a central focus. Companies outside the aerospace industry needed a trusted adviser to guide them in designing experiments and choosing space hardware — and they needed a champion to represent their interests when U.S. and European governments planned future space efforts.

INTOSPACE has 94 company members from nine countries. They are looking into both research and actual on-orbit manufacturing ideas. Its three automakers — BMW, Daimler-Benz and Volkswagen — would benefit from physical chemistry research into corrosion, for example. Understanding the physics of rust through microgravity experiments could improve the quality of their cars. They also would benefit from polymer/plastics research similar to that planned by GM and 3M in the United States.

Other INTOSPACE firms deal in chemicals, steel, aluminum, electronics, and biotechnology. Banks and insurance companies have joined to keep in touch with their customers, as have several aerospace firms.

Hans Hoffman, the charismatic head of INTOSPACE, says interest in space research far outruns the available facilities. Companies have proposed twice as many experiments for the next German Spacelab, the D2 mission, as it can hold. He's fairly peeved that NASA won't launch the D2 Spacelab until late 1991, six years after the initial D1 mission. Research that must wait six years between experiments can ruin the careers of its sponsors.

NASA will launch a half-dozen communications and navigation satellites ahead of the D2 Spacelab, even though several could be cleared from the shuttle schedule by launching on expendable rockets. The Spacelab hasn't any alternative — the shuttle is the only vehicle capable of carrying the Spacelab and its scientists.

In the meantime, INTOSPACE can fall back on small, unattended experiments in the shuttle. The shuttle experiments can be carried on the popular and crowded middeck area or in a middeck extender called Spacehab that an American firm is developing (see Chapter 11).

Some experiments are possible on sounding rockets, which are very small boosters that never reach orbit. As the sounding rocket coasts through the top of its arc and starts to fall back down, the payload gets about six minutes of microgravity. The "quality" of the microgravity in a sounding rocket is very high—there aren't any astronauts shuffling around creating vibrations. On the other hand, the experiment has to be completed in under six minutes and can't require very much electrical power.

A firm called Microgravity GmbH in Munich has taken over the German government's TEXUS sounding rocket program—a sharp contrast to the U.S. situation. NASA considered turning its small Scout rockets over to private hands, but the agency decided there simply isn't enough American interest to support a private operation.

INTOSPACE also has employed communist rockets, as described in Chapter 2. Many European experimenters now plan to use the Soviet space station *Mir* as well, because American flight opportunities are so scarce.

AMERICAN RESEARCH CENTERS

The only real center for microgravity research in the United States until recently was NASA's Marshall Space Flight Center near Huntsville, Alabama. Marshall handed out relatively small grants to universities like MIT to keep academic interest in space alive during the preshuttle era.

With the debut of the space station program in 1984, NASA began pushing out more grants for microgravity work. The funding created users for the space station, thus improving chances that Congress would fully support the program. Aerospace companies also cranked up microgravity projects. They hoped that expertise in space station uses would give them an edge in winning the contracts to build space station modules.

Consulting and accounting firms also dived into the new market. They teamed up with various aerospace firms working on space station contracts and toured the country trying to drum up commercial interest in the space station. NASA hired the consulting firm of Booz, Allen & Hamilton and the accounting firm of Coopers and Lybrand to work directly for it on creating new space station users. Booz Allen and Coopers developed sample microgravity ideas tailored to specific industries. Their consultants visited dozens of target companies and used these examples to spur interest.

In 1985, NASA shifted its dollars from traveling consultants to university-based centers for commercial development of space. NASA awarded up to $1 million each to 16 universities and nonprofit research centers. The winners were chosen partly on how many industrial partners they lined up. The goal was the same as with the consultants: Draw mainstream manufacturers into space R&D so the space station would have a growing pool of potential customers.

In addition to the University of Alabama-Birmingham's Macromolecular center mentioned previously, the University of Alabama-Huntsville also won a grant, to support development of optical computer crystals, protective coatings, and other areas. Boeing Aerospace, IBM, Deere and Company, Wyle Laboratories, and several aerospace firms are helping to pay the bills.

Vanderbilt University won NASA cash for a commercial center for metallurgical processing in space. It will use containerless processing, directional solidification and vacuum solid state processing. The center has attracted Alcoa, Armco, Allied Signal, Englehard Corp., General Electric, GM, General Telephone & Telegraph (GTE), and others. While some may hope for high-value alloys they could produce in space, the more likely outcome is greatly improved understanding of their current Earth-bound factories.

Battelle Columbus Laboratories in Columbus, Ohio, won NASA backing for a commercial center with broad ambitions. Battelle will investigate multiphase material processing in metals and alloys, glass, ceramics, electronic and optical materials, and polymers. University members include Akron University, Case Western, Clarkson, Cleveland State, Ohio State, and Washington State. Industrial partners range all over the lot, from Amoco Chemicals and PPG Industries to the small semiconductor firm II-VI Inc.

The Institute for Technology Development at the National Space Technology Laboratories in Hancock, Mississippi, was picked as a remote sensing center. Partners range from Montana State University to the International Paper Company, which would use remote sensing to manage its forest properties. Space cameras can spot diseased trees and insect blights more efficiently than ground-based surveys.

In 1986, NASA gave grants to four more universities for commercial development centers. The University of Wisconsin at Madison is working on a triple play with its money — a trademarked trio called Astrorobotics, Astroculture, and Astrofuel. Astrobotics will develop automation and robot technology to aid human expansion into space. People are extremely expensive to deliver to orbit and very costly to maintain once they're there. And when outside work needs to be done, humans encased in space suits just aren't very efficient. Robots designed to work in a vacuum will be essential to leveraging the human presence in space.

The Astroculture program is tackling the problem of supplying food and eliminating waste through on-orbit agriculture. A closed-cycle life support system — one that doesn't need Earth supplies to keep functioning — will slash the cost of operating space habitats.

The Astrofuel program is designed to gather Helium-3 from lunar and planetary sources. Helium-3 could make nuclear fusion an economical source of electricity on Earth. The lunar soil is believed to be a gold mine of helium-3, and the university's robots could be trained to harvest it for shipment back to Earth.

The University of Wisconsin has an odd lot of corporate collaborators, from Johnson Controls to Snap-On-Tools Corporation.

Ohio State's Center for Mapping is to develop more remote sensing applications. One of the long-running problems with remote sensing is that millions of images are collected but are never used. Some are wasted because they reach users too late to be useful, and others aren't used because too much effort is required to extract information from the pictures.

Ohio State will develop ways to quickly merge remotely sensed information with standard geographical data bases. It also will seek new programs that extract information from photos. An example might be a program that automatically turns a photo into a map of drought areas or that automatically highlights the best areas for mineral exploration.

The University of Houston staked out a unique position—it's the only center devoted to putting the vacuum of space to work. The vacuum is an ultraclean environment that will allow creation of perfectly pure materials and devices. The university's Space Vacuum Epitaxy Center is building machines to deposit atoms of the desired material layer by layer. To get the best vacuum, the work will be done behind a wake shield shaped like a satellite dish antenna. The shield brushes aside any stray oxygen and other atoms that still remain even in low Earth orbit.

The Space Vacuum Epitaxy Center may create microelectronics, optoelectronics, high-temperature superconductors, microwave electronics and ultrasensitive sensors. The center's industrial partners are putting in almost a million dollars in cash and services. The partners include AT&T Bell Labs, Perkin-Elmer Corporation, Electro-Optek Corporation, Rockwell International, and the U.S. Army.

Clarkson University in Potsdam, New York, put together a broad academic coalition for crystal research. The center will run on a budget of more than $2 million put up by NASA, 12 corporations, and the states of New York and Florida.

Clarkson and the University of Florida will develop cadmium telluride and gallium arsenide for integrated circuits, microwave devices, and infrared sensors. Rensselaer Polytechnic Institute's role is to create radiation detectors by boiling mercury halides and letting the gas condense on a target. Alabama A&M is growing infrared detectors and electrooptic crystals in liquid solutions. Worcester Polytechnic Institute will continue its research on zeolite crystals that could be used for catalysis, biotechnology separations, and radiation waste disposal.

Corporate sponsors range from Spacehab (see Chapter 10) to Westinghouse and Grumman Corporation.

NASA approved seven more Commercial Development Centers in 1987, focusing on space power, space propulsion, and biology. A complete list of the centers can be found in the back of the book.

THE JAPANESE INITIATIVE

In Japan, announcement of NASA's space station released a burst of interest in microgravity. The Science and Technology Agency, which oversees the Japanese space program, began holding monthly meetings for companies interested in the space station and mircrogravity—with more than 600 participants at every meeting. The government set up a Space Utilization Promotion Center in 1986 endowed by grants from 42 corporations, and five major Japanese trading companies formed study groups with their subsidiaries and suppliers.

The Japanese aren't exploring microgravity from idle curiosity. They smell serious money. One of the study groups—with Mitsui, Toshiba, and 70 other firms—expects space manufacturing between 1992 and 2010 could generate $15 billion in profits. The Mitsui group is interested in separation of living cells (the McDonnell Douglas specialty), new electronic materials, and super heat-proof alloys.

Other heavyweights in microgravity research are Sumitomo group, Nippon Electric, Nissan Motors, Hitachi, Kawasaki, Kyocera, Kobe Steel, C. Itoh, Mitsubishi, and Marubeni.

The government also established a Space Technology Corporation in 1986 to spearhead work in new semiconductors. The corporation has leased space on the German D2 Spacelab, paid for by its six member companies and the Ministry of International Trade and Industry (MITI) and the Ministry of Posts and Telecommunications.

The Japanese began zero-gravity experiments in 1980 with a series of sounding rocket flights. They concentrated on new semiconductors, an obvious target for a nation so dependent on advanced electronics. But the flights also tested new composite materials. The composites included aluminum strengthened with carbon fibers and glass strengthened with diamond particles.

Only a few Japanese experiments flew on the shuttle before the *Challenger* disaster. These were student-level demonstrations sponsored by the Asahi newspaper/broadcasting group. The projects made artificial snowflakes and shot metal balls at water drops. The water drops showed how the surface tension of a liquid is very important in space. Metal balls fired at slow speeds (under 30 centimeters per second) hit the water drop and stuck to its surface. Medium-speed balls entered the water droplet but couldn't break out the other side. Fast balls (90 centimeters per second) streaked right through the water drop — and on their way out, they were followed by a baby water drop the same size as the metal ball.

The target practice demonstrated that surface tension in space will tend to form liquids into balls that fight intrusions of foreign objects. Scientists are finding this an extremely vexing problem when pumping fluids from one part of an experiment to another. The liquids try to clump up into balls. That ruins the smooth, even flow of fluids required in some experiments.

The first large-scale Japanese use of the shuttle will come in 1991 with a Spacelab flight of 22 materials-processing experiments. Three Japanese scientists are in training now to become the payload specialist on that flight. When the U.S. space station is on orbit, the Japanese plan to add their own module for materials processing, remote sensing of the Earth, new communications technology, and life science experiments. Potential users have submitted more than 300 proposals for space station tests.

THE SOVIET SCIENCE PUZZLE

The Soviet Union launched the world's first space station in 1971. They followed up with six more. Since 1977, the Soviets have enjoyed almost continuous access to space station resources. The United States, by contrast, had a single year of space station operations during the 1973 *Skylab* mission. The Soviet Union's clear lead in space stations should translate into a clear advantage in materials processing, but very little has been published to document their lead.

In 1986, for example, the Soviet Union gave a review of materials-processing work before the International Astronautical Federation in Innsbruck, Austria. The presentation by L. L. Regel of the Space Research Institute said the Soviets had learned how to weld in space and how to repair thin-film coatings that protect their space stations. She acknowledged that Soviet scientists had completed experiments in crystal growing and metal alloys. But her paper gave very few details on any of the Soviet work. Most of her talk covered results from Western research done on the first American Spacelab mission.

Have the Soviets failed to take full advantage of their space stations, or aren't they allowed to talk about that research? Probably both explanations are true.

Censorship of scientific papers inside the Soviet Union inevitably has slowed the pace of research there. Even when Soviet results are published, they simply present the facts—scientists don't speculate on possible applications. Unleashing the imagination is half the battle with microgravity research, so the sterile Soviet style doesn't help other Soviet scientists grasp the ramifications of new research.

Soviet materials science also seems to be missing the backup tools available to Western researchers. If a Soviet scientist lacks the clout to get an experiment on their space station, he or she has nowhere else to turn. Americans waiting for space shuttle access can fall back on sounding rockets and drop towers. Regel, who gave the prime Soviet paper at the IAF Congress, apparently had never flown her own experiments on the Soviet space station. She had to use centrifuges to create multiple-gravity situations, and then reason backwards to what impact zero gravity should have.

The Soviet space stations themselves have another limit: puny electrical power. Even their latest station, the third-generation *Mir* orbital complex, has only 20 kilowatts of electricity. That's about equal to the peak power drain of one American all-electric house. (The U.S. station will have 75 kilowatts when it finally gets into operation.) Large-scale materials processing would take more electrical power than the Soviets have available. Regel said the largest crystal-growing experiment planned for *Mir* would use about 500 watts.

Western experts note that the solar panels on the *Mir* complex are still made from old-style silicon chips. They speculate that the Soviets would have switched to more efficient gallium arsenide solar panels if their previous space station work on gallium arsenide had made significant progress.

So the outlook on Soviet materials processing is murky. They haven't demonstrated a tremendous lead, but what's visible certainly isn't the full story.

The worldwide outlook is definitely more upbeat. Several hundred companies around the world have started to include microgravity in their research plans. Activity on this scale is bound to produce innovations in both Earth-bound and space applications when the U.S. space station opens for business in 1997.

Spotting Earthly Treasures
with Orbital Cameras   9

Military spy satellites were the first to exploit the "remote sensing" of Earth. Their reconnaissance photos also received the first extensive digital enhancement, as U.S. intelligence agencies fed them into the world's most advanced computers to bring out details on missile sites and troop movements.

Today that image technology is available to the smallest civilian companies and even well-heeled hobbyists. Anyone can buy satellite photos and enhance them with a personal computer. A burst of new ways to use space photos will result, now that individual experimenters can use the tools once reserved for intelligence analysts.

Civilians now use satellite photos to investigate rich mineral deposits, good ocean fishing areas, worldwide crop trends, or the ruined Soviet nuclear reactor at Chernobyl. The photos are available from weather satellites and specialized remote sensing satellites launched by the United States, France, and the Soviet Union.

Remote sensing is a very close cousin to the comsat industry. They both exploit space for information handling rather than the manufacture of orbital products, getting around the cost-per-pound drawback of current launch systems.

The similarity to comsats indicates that remote sensing should become a thriving space industry. The Reagan administration tried to invigorate remote sensing in 1983 by offering to sell government satellites to industry.

The sale turned out to be premature. Paying customers for remote sensing were scarce, and most of them were federal agencies. The "sale" became a fight over subsidies. The new private owner desperately needs subsidies to survive, and federal budget officers are determined to squash them.

Subsidies naturally result because the U.S. Landsat remote sensing system was built to serve government needs, not the private marketplace. Federal agencies buy half the Landsat photos and account for three quarters of the revenue. The Central Intelligence Agency (CIA), other U.S. spy services, and the Foreign Agricultural Service pay premiums to get priority use of the satellites.

These government users often need broad summary data. What's the Soviet wheat harvest expected to be this year? Will a freak freeze in Brazil's coffee region send prices soaring? Government computers crank out projections, and unless they are classified, everyone shares the results at no cost. The system is geared to producing information of broad public interest that usually is distributed for free.

Under these conditions, entrepreneurs trying to turn Landsat into a profit-making venture serving private interests have real handicaps. The photos themselves aren't very detailed — interstate highways show up, for example, but many major city streets are barely visible. People seeking information about a particular site therefore must put up with vague outlines and fuzzy lumps. And the government has decreed that all photos must be made available to the public — no one can buy exclusive photos and develop truly private information.

THE TREND TOWARD "PERSONAL" REMOTE SENSING

A strong market for space images hinges on sharper pictures and the computer power to manipulate them effectively. Then people will be able to use satellite photos to track their own small, specific interests.

The market would be even stronger if U.S. policy allowed "private" space photos, but that would create a worldwide political uproar. Third World nations already fear that advanced countries use public remote sensing to spot their mineral wealth. Allowing corporations to own private space photos — images denied to the target country — would create a storm of protest from developing nations.

Even without private ownership of space images, the next generation of satellites will have a crucial advantage with their sharper photos. They will be able to provide images of individual factory sites and farms. A farmer, for example, will put the space photo into his personal computer and find which sections need irrigation or an application of fertilizer.

Farmers will save money by irrigating or fertilizing just the specific spots with problems, rather than an entire field. The savings would justify the cost of buying the space image. Providing this kind of help directly to individual users will be the key to any successful remote sensing operation.

The two requirements for this system — sharper pictures and better computers — are coming. The French SPOT system is shaking things up, providing photos that can pick out objects only 5 to 10 meters wide. That's far better than the older U.S. Landsat system, which the U.S. military held back to only 30-meter and 80-meter resolution. Future SPOT satellites will be even sharper, offering one- to five-meter resolution.

The French initiative probably will force U.S. officials to loosen the national security restraints on American satellites. The change won't come easily—the technology is so secret that U.S. officials in the mid-1980s refused to say in public how sharp a civilian system can legally be. But, eventually, civilians will be able to buy space photos that can count the cars in a parking lot or provide farm-specific information.

The needed computer power is coming, too, in the form of desktop computers using advanced chips. Advanced personal computers are essential because satellite image software requires massive computing power. A single image may take up 25 megabytes of hard disk space. That's a tremendous amount of data for a computer to store, much less actually manipulate simultaneously to create an image. As a yardstick, when the first hard disk for personal computers debuted in 1980, it held only five megabytes and cost $1,500. Now a hard disk four times larger and three times faster costs only $250.

Personal computers in the early 1980s were so puny that anyone trying to enhance satellite images had to use minicomputers or mainframes.

But today's personal computers using the Intel 80386 or Motorola 68030 chips rival the $200,000 minicomputers sold just a few years ago, or the $3 million IBM mainframes sold in the 1970s. And personal computers now have the hard disk storage needed to cope with satellite images—120-megabyte drives are common, and drives with more than a thousand megabytes are available.

Computer programs bring out the details needed to identify a good oil drilling spot, or they organize the data needed to make crop forecasts. The software for each purpose often is unique, which creates room for dozens of small firms to create valuable remote sensing programs tailored for a specific market.

A firm in Mountain View, California, has put all the elements together in a package deal. For $30,000 to $45,000, Terra Mar will provide the computer, high-capacity disk and tape drives, high-resolution graphics monitor, and the special software needed to turn satellite photos into useful information. The system provides analysis powers that used to be the domain of the CIA and National Security Agency.

COMPUTERS SHARPEN THE VISION

Satellites transmit their photos in digital format, so computers are needed simply to decode the signals. But computers do much more, allowing experts to probe for information hidden to the naked eye. The first step often is contrast enhancement. The satellite camera sends back an image as a series of numbers, with each number representing a point in a scene. The size of the number shows how bright the point is, usually on a scale from 0 to 255. An entire grid of these points creates the total image.

Two neighboring points may have very similar values. For example, they may have brightness values of 190 and 192. The human eye probably won't notice the difference. The computer software can highlight such variations by doubling or tripling any small difference between groups of neighboring points. The brighter point would be made brighter, or the darker point would be made darker. Points that already have a large difference from their neighbors would be left alone.

The next step in photo enhancement is to create a color image. Remote sensing and weather satellites usually can't supply a normal true-color photo made from separate red, blue, and green versions of a scene. Their cameras instead take six or seven versions of a scene. Some images may come close to the standard red-blue-green split, but others will be taken in the infrared part of the spectrum. Infrared light is a longer wavelength that the human eye can't register, but it's very useful in revealing variations in plants and minerals that don't show up in visible light.

Each of the six or seven versions of a scene each is transmitted back to Earth as a simple black-and-white photo. The shades of gray are created from numbers (from 0 to 255) sent back by the satellite cameras. In the "blue" version, the "bluest" parts will have higher numbers and the "not blue" parts will have lower numbers. The same goes for each of the infrared versions of the scene. (There may be three or four infrared versions, each taken at a different wavelength, or "band.")

The computer software can arbitrarily assign a color to any of the versions. One infrared band can be given the color green, another band can be assigned red, and a third band could be made blue. The assignments create a false-color picture that emphasizes the variations seen below. One band labeled with green might be good at identifying which fields are planted in corn, and the band labeled with red might be good at pinpointing infestations of caterpillars. The blue band could be the one best at identifying water-soaked soil. Combining the three bands creates a false-color image where corn fields are clearly seen in green, with insect problems highlighted in red and poor drainage highlighted in blue.

Researchers have come up with an ingenious extension of this color-image trick. They take the numbers representing one version of a scene and subtract them from the numbers of a different version. This technique can pick out very subtle variations in a scene. Consider how the sunlight hits a rolling terrain. One side of a valley gets the sun at a low angle and looks darker, even though the crops or minerals are the same on both sides of the valley. Subtracting one band from the other cancels out the effect of sun and shade. What remains are any differences in how light is reflected in the two bands—healthy vegetation may reflect equally in both bands, but diseased plants may look dark in one of the bands and thus stand out.

Imagine a forested valley. Trees on the bright side of the valley may be represented by the number 200 in both bands and trees in the dark side of the valley may show up as 120 in both bands. Subtracting one band from the other gives a clear base level of 0 for both sides of the valley.

Now consider a patch of gypsy moths in one part of the valley. They disrupt the normal difference in how the bands reflect light. Trees infested with gypsy moths on the bright side may still read 200 in the first band but only 175 in the second band. The difference between bands suddenly shoots up to 25 for the infested trees. The 25s will stand out in a sea of 0s. The dip in the second band, from a normal 200 down to 175, might be missed completely using normal techniques. The slight dip from gypsy moths would have been mixed up with a vast jumble of dips caused by shadows, trees of different ages, and a host of other causes. The two-band subtraction method cancels out any variation that affects both bands equally. The differences that remain scream out for attention and often show some significant change in the crops or minerals below.

The U.S. Landsat system sends back seven bands on each scene. The subtraction method can be used with several pairs in creating a false-color picture. One subtraction can control the hue, another controls saturation, and a third controls the intensity. The result is often a brilliant color image showing features that are completely invisible to the naked eye.

A more sophisticated approach than subtraction is division of one band by another. In the example above, one band registered a 175 while the other had a 200, for a difference of 25. But what if the two bands had registered at 25 and 50? The absolute difference is still only 25, even though the jump from 25 to 50 is a tremendous relative change. It's a 100 percent increase. The shift from 175 to 200 is a mere 14 percent increase and probably isn't nearly as important to spot as the doubling from 25 to 50. Division of one band by another lets this relative importance come through clearly. If the

two bands "see" the land the same way, division will produce an ocean of
ones. Dividing 200 by 175 will produce a 1.14—not so very different from
all the ones around it. But dividing the 50 by 25 will produce a 2 that stands
out sharply.

THE WINTER WHEAT INVESTIGATION

A NASA-sponsored center is using this multi-band technique to spot
winter wheat fields that may produce poor yields. A fertilizer company is
helping with the experiment—it wants to survey its market area from space
for farmers who ought to apply more fertilizer.

The project is supervised by the Institute for Technology Development
(ITD) near Jackson, Mississippi. Working with Murray State University and
the Hutson Fertilizer Company, ITD is testing the idea in western Kentucky.
The spring harvest there in 1986 was very spotty, with yields ranging from 15
bushels per acre to more than 50 bushels. Many of the stunted fields later
were found to be poor in nitrogen. The fertilizer company hopes that satellite surveys can identify the weak fields in time for an emergency dose of
nitrogen to restore yields.

The project has two questions to answer: Can Hutson and its scientific
advisers identify the fields that need nitrogen? And can Hutson get access
to the Landsat photos quickly enough so that the extra nitrogen actually will
make a difference in yields?

The 1986 work succeeded on the first question. Landsat photos taken
near the end of the growing season were able to spot the fields with a poor
harvest. Each field was farmed by the same operator, had identical soils,
and received equal fertilization early in the season. The best field had a
distinctive peak in band 4 of the space photos, compared to the more
spread-out reflections from the malnourished field. ITD researchers say this
needs to be refined into a prediction of the precise yield in bushels. Then
farmers can calculate whether the cost of applying more nitrogen will be
profitable.

The second question—actually getting Landsat photos quickly—is yet to
be answered. When the federal government ran the program, it didn't have
enough money to process all the photos taken by the Landsat cameras. The
ones that were processed reached users months after they were taken. Now
the Earth Observation Satellite Company (EOSAT) is running the Landsat
cameras, and the lag time is three or four weeks. EOSAT and federal
budget officials are battling over government subsidy levels, however, and

EOSAT began laying off workers when its subsidy demands weren't met.

This bickering over Landsat funding makes customers very uneasy — there's no guarantee that reliable, prompt service will continue. Investing in analysis software or a service business based on Landsat photos thus may be a real gamble until the government finally decides how much support it will give remote sensing.

PLAYING THE COMMODITIES MARKETS

Of course, not all money-making ideas hinge on detailed farm-by-farm photos. Broad summaries of world crop conditions can be valuable if you are a commodities speculator or grain trader. Crop prices make sharp swings whenever the official U.S. government forecasts are revised. A private system that could anticipate these changes before they are announced could be extremely lucrative.

The Agriculture Department ran a pilot program called STARS that used satellite data for crop predictions. Analysts employed Landsat photos to classify fields into various crops and to estimate their yields. The standard Agriculture Department forecasts are built from status reports sent in by thousands of county agents monitoring crops in their area. The STARS estimates worked fairly well but generated no enthusiasm from top-level officials who controlled the budget. Funding for the effort eventually died out.

Nothing prevents private interests from continuing the STARS program. In fact, powerful personal computers make data handling far less expensive than it was when the Agriculture Department tried the STARS program in the early 1980s. Private forecasts can be built around data sent back from weather satellites. Weather satellites have several advantages over Landsats. The weather photos are available twice daily, instead of every 18 days for Landsat images. The weather photos also are less detailed, which can be an advantage in cutting down the amount of data the computer has to manipulate.

Polar-orbiting weather satellites carry an instrument called the advanced very high resolution radiometer (AVHRR). Like most space cameras, it takes images in several spectral bands. Two of them are especially sensitive to the chlorophyll and mesophyll in vegetation. The Commerce Department creates a weekly Global Vegetation Index based on these two bands. The index shows whether the vegetation is thriving or suffering from drought or insects.

The AVHRR images aren't precise enough to identify which areas are planted in corn or wheat and which are simply growing grass or trees. Thus, they can show how well the vegetation in Argentina is doing this year but not how many acres had been planted in wheat. A private forecasting system could get around this by using Landsat images a few times a year to map the land under cultivation for major crops. With the coordinates for major crop areas in its memory, a computer then could use the weather satellite photos to spot major trends in each crop.

The agility of a personal computer in scanning AVHRR data is graphically shown in a system developed by Richard Bolton and Michael Rega of the Commerce Department's National Environmental Satellite, Data and Information Service (NESDIS). They put a graphics card into a simple IBM PC clone with a small ten-megabyte disk, and created software to handle AVHRR images. The user can feed an entire continent's data into the PC with five or six floppy disks. (The disks must be created first by a VAX minicomputer at NESDIS.) The color PC screen shows the full continent, and the user can zoom the image to show a detailed section. The colors appear natural with white clouds, red dry soil, brown wet soil, and lush vegetation in bright green or blue. Sparse or dried-out vegetation shows up in orange, yellow, and pale green.

The NESDIS system is aimed at developing countries that don't have thousands of county agents to watch over crop trends like the United States does. With improvements—like coordination with Landsat surveys—the NESDIS system might be an excellent basis for a private crop forecasting system. The owner of such a system could profit by buying grain or futures in advance of official reductions in crop forecasts, or by selling them before release of increased projections.

The Earth Satellite Corporation in Chevy Chase, Maryland, already produces some private forecasts under the trademark "CROPCASTS." The forecasts only cover 12 countries, however.

THE ORBITAL TAX ASSESSOR

Big Brother may soon start watching taxpayers as well as troops from orbit, toting up property tax bills based on satellite reconnaissance. State tax authorities are being offered orbital photos to determine the value of rural land, in states that tax land based on its use. Land laying idle is taxed at a lower rate than fields growing soybeans, for example. ITD and two small firms are developing the hardware and software to track this land use by satellite. State tax assessors will be able to churn out bills based on the acreage and yield of every square yard owned by each taxpayer.

The companies trying to give tax assessors a bird's eye view are Geo-Information Systems of Starkville, Mississippi, and GeoDecisions of State College, Pennsylvania. Both are start-up ventures spun off by local universities. They recognize that figuring the farm property tax is no simple matter. The tax varies with the soil type and whether the land is growing a crop. The boundaries of various soil types will zig and zag, so a field of corn may have three or four different tax rates applied to it. The companies hope to match orbital photos with a stored digital map of soil types and compute the tax bills automatically.

Another ITD idea is automated site selection. Satellite photos of land cover would be combined with digital maps of terrain, soil types, utilities, and roads. Each factor is rated, and the combined ratings produce a map showing the best sites. ITD says three companies are interested in taking the idea into production.

PROSPECTING FOR BURIED RICHES

Satellite photos can help spot oil fields and ore deposits, using hints that appear on the surface. One hint for oil is the appearance of lines or folds on the Earth's surface. These ripples are called lineaments. They show where the Earth's crust has been fractured or folded over. The fractures seal off an underground region so that any oil formed by decay of ancient vegetation will be trapped.

Energy companies are working on computer programs that automatically interpret satellite photos, looking for lineaments. The programs work almost in a connect-the-dots fashion. The cracks or folds are only partially visible—some portions may be hidden by plants, landslides or human construction, so the programs draw lines to connect the visible sections together.

Looking for mineral deposits is possible as well. No surefire method has been discovered yet, so prospectors must rely on clues rather than plain evidence of an ore body. One clue to buried iron ore is a change in the color of vegetation growing above it. Tiny amounts of the iron leach out and reach the surface, stunting the growth of any plant cover.

Weather satellites equipped with infrared cameras also can help spot large ore deposits by the way they retain heat. The infrared images can measure the temperature of the ground during the day and at night. Areas that cool off more slowly at night stand out—they are the denser rocks that have retained the most heat. The thermal inertia of these rocks may point the way to a large metal ore deposit.

FISHING FROM SPACE

Since the dawn of humankind, fishermen have realized that fish must be tracked down to their favorite spots. Now some commercial fishermen are using satellite photos to go directly to ocean areas with the best prospects.

None of the satellites actually can identify schools of fish. Commercial fishing fleets instead go for places rich in fish food, and hope the fish have found them as well. The first space fishing maps, issued by the Commerce Department, used temperature readings taken by polar-orbiting weather satellites. Cold spots in the ocean meant that deep water, rich in nutrients, was welling to the surface. Albacore tuna and salmon fishermen usually had good luck along the boundaries of these food-rich areas.

The latest technique looks at the color of the ocean water using a Coastal Zone Color Scanner aboard the Nimbus-7 weather satellite. This scanner detects areas rich in chlorophyll and phytoplankton. Heavy plankton growth, the basic plant in the ocean food chain, points to good fishing. The plankton may be spurred by an upwelling of cold water, or it may have other causes. So tracking plankton is more accurate than tracking cold/warm boundaries — it's the plankton that attracts the tuna and salmon.

SPOTTING MOSQUITOES FROM ORBIT

World health experts say mosquitoes and malaria are on the comeback trail. Some mosquito strains have become resistant to pesticides, and some malaria parasites can no longer be killed by medicines like quinine. Governments are looking for new ways to regain the upper hand.

One answer is pinpoint targeting of malaria outbreaks with satellite photos. Space images can speed health workers to the precise areas with the worst pattern of rainfall, standing water, and temperature. They can drain stagnant surface water and apply heavy doses of pesticide to the critical area. A "surgical strike" like this avoids the cost and environmental damage of spraying pesticides over wide areas.

NASA is working on the problem through its Ames Research Center in Mountain View, California. The first step, taken in 1987, is to see if NASA can predict mosquito outbreaks in California rice fields. The mosquitoes there are malaria free but are the same type that transmit malaria elsewhere.

Speed is the key goal, because the fast-breeding mosquitoes multiply based on day-to-day changes in rainfall and temperature.

If perfected to work in the tropics, NASA will give the technology to national governments or international agencies. The World Health Organization says malaria afflicts 250 million people and is the leading cause of disability in the tropics.

AN ORBITAL MONITOR FOR TOXIC WASTES

Satellite cameras are very good at detecting subtle changes in plant growth and ground color. This may lead to a new business in policing hazardous waste dumps, by watching the surrounding land to see if wastes are leaking off the sites.

The core "technology" of a hazardous waste dump is containment. The site is supposed to trap the wastes and keep them immobile. Some dumps do this by putting down a layer of clay, forming a barrier between the wastes and the rest of the world. Often the clay barrier fails—a crack develops from the settling of the ground underneath, for example—and the poisons begin to seep out into the local soil and water supplies.

A remote sensing satellite can watch over dump sites, looking for signs of seepage. Plants growing downhill from the site may show the first signs of a leak. The pattern of stunted or discolored vegetation may be visible only from space.

Industrial plants also could be policed from space. Some river polluters have been known to dump their wastes at night, when the strange colors in the water will be hard to see. Infrared monitors in space could easily pick up this kind of evasion, because the waste is almost always warmer than the river. Dumping at night would only make this temperature difference stand out more in the satellite photos.

In a few years, governments may require dump sites and industrial plants to post bonds that guarantee constant orbital monitoring. Private firms operating the surveillance systems would sell the bonds. It could operate almost like Orkin's termite service—Orkin guarantees to keep your home termite free, or it pays for the eradication and repairs. A space monitor firm might guarantee to state and federal governments that it will discover every pollution incident or pay for the cost of cleanup itself if the breach is caught first by someone else.

A Fresh Point of View

Most space photos show the view from directly overhead, like a map. New computer techniques can transform overhead shots into side views, as if the camera were at ground level. Architects or geologists scouting a site can use the computer to take a walking tour.

The perspective can be shifted to airplane level, too, showing the perspective that would been seen by a fighter or bomber pilot. Military planners will use this to create simulated missions over enemy territory, showing pilots what they'd see out their cockpit windows, using only a satellite photo for data.

THE NEW TECHNOLOGIES

The satellite cameras in orbit today are somewhat crude at analyzing the Earth by its reflected light. The best cameras only break the light down into six or seven bands. Laboratory spectrometers are able to break reflected light into 200 or more bands. With 200 different views of the same object, a laboratory spectrometer usually can identify its composition with precision. That's often not possible using just the six or seven views available from satellite photos. The satellite photo will spot when the ground below changes to something new, but it may not reveal whether the change is caused by an ore deposit or by an outcrop of granite rock, for example.

Taking photos in several hundred bands simultaneously presents one big problem: bloated data. Each version must be transmitted back to Earth digitally and then, somehow, stored economically. Users find that hard enough now with only seven bands. The challenge is to find automatic methods of data reduction or compression.

One method is to create computer programs that throw away any information that is repetitive. For example, the image may have a long stretch of grass, perhaps 100 consecutive points where the reflected light doesn't vary until a tree shows up. Instead of storing a hundred numbers that are all the same, the data base can store two numbers: one giving the intensity of the grass reflection, and another showing that this number should be repeated 100 times.

Another method is to learn exactly the telltale signs of the target, such as a gold deposit. Then the program can throw away any data that doesn't fall within the ranges of a gold strike. Information about the surrounding trees, lakes, or crops is simply flushed if it doesn't directly relate to finding the gold.

The first orbiting spectrometer is scheduled for flight on the space shuttle in the early 1990s. The Shuttle Imaging Spectrometer Experiment (SISEX) is to monitor 128 bands. To keep the data manageable, the SISEX photos will cover a smaller ground area than Landsat photos — only 12 kilometers wide for SISEX compared to 85 kilometers for Landsat. Sometime in the mid-1990s, NASA hopes to put a more advanced version into regular service in polar orbit. This High Resolution Imaging Spectrometei (HIRIS) would sweep a 50-kilometer-wide swath.

Another major improvement in remote sensing is the use of radar waves. Several shuttle flights have carried "imaging radars" that bounce radar signals down to the Earth and back. The radar waves can penetrate the surface to reveal buried structures such as old river beds and fault lines. Geologists can use these hidden clues to search for oil and minerals.

Radar images aren't affected by clouds and can be taken 24 hours a day. This versatility is especially important for military surveillance systems that must be able to see through clouds and to see at night. Radar also will help civilian applications in regions with chronic cloud cover.

Computer systems now on the market can combine space data from several sources. Radar images can be added to Landsat, SPOT, and weather photos for a master view of the Earth. The profile can catalog the world's resources and problems more completely than an army of Earth-bound surveyors and explorers.

Prime Real Estate:
The New Space Stations   10

Space stations will be one of the industrial battlegrounds of the twenty-first century. Their owners will seek wealth from two basic operations. Some stations will be truck stops to repair and refuel space vehicles, and others will be industrial parks leased out for research and manufacturing. By the year 2005, there likely will be a half-dozen private, NASA, and military space stations in service.

The truck-stop space station will refuel communications and reconnaissance satellites, in addition to spaceships. Comsats burn fuel to stay in the correct orbit, and some spy satellites use fuel to change orbit on command to scout out activity below. Often satellites will need repair work, so the crew at a truck-stop station will replace burned-out tubes and fix damaged solar panels.

This station also will resemble a river port, where inland barges transfer their cargoes to ocean carriers. A similar operation will be done in space, as cargoes are delivered to the station for later boosts into higher orbits. Communications satellites will arrive at the station on Earth shuttles derived from the National Aerospace Plane. They'll get a diagnostic check for launch damages and be put on orbital transfer vehicles (OTVs). These are small, unmanned space tugboats. The OTVs will carry the satellites up to geosynchronous orbit and then return to the space station for another flight. This robot taxi service will save money because the OTV is stored on orbit and reused constantly. The current method requires boosting a fresh upper stage into space with each satellite. Not only does the customer have to buy a new upper stage for every satellite, but the weight of the upper stage cuts into the payload that the shuttle or expendable rocket can carry.

The OTVs may cut costs further by using fuel created in space, instead of fuel expensively hauled up from the surface. Waste water from the space station, for example, could be turned into oxygen and hydrogen with electricity from the station's solar panels. Oxygen also could be extracted from Moon soil and delivered to the station, at a fraction of the cost of boosting it all the way from the Earth's surface. (Six times more oxygen than hydrogen is needed for the typical rocket engine.) The station also could beam electrical power to the OTV in flight. The energy could run a laser engine like those described in Chapter 4 or a standard electrical ion engine like those that have powered deep space probes.

A truck-stop space station could use tethers as a no-fuel way of boosting and retrieving the OTVs. As explained in Chapter 4, tethers are long cables. A space station can reel an OTV out on a tether, letting it go up to a higher orbit as the space station moves down. The OTV will move up a lot because it's light, and the station will move down only slightly because it's massive. When released, the OTV will be flung to an orbit seven times higher than the length of the tether. Cables more than 300 miles long are feasible, so the seven-times boost will put the OTV into an orbit thousands of miles higher without burning a drop of fuel.

The station would become an energy bank—flights going up would make withdrawals and traffic going down would make deposits. Down traffic might include returning OTVs, shipments of oxygen or metals from the Moon, or shuttles returning to Earth. Tethers would allow them to donate some of their energy to the space station, moving it to a slightly higher orbit. Flights going up—such as OTVs headed for geosynchronous orbit—would take out energy, and the station would drift to a slightly lower orbit.

A fleet of efficient OTVs would transform existing launch systems. The Titan IV can haul 10,000 pounds to geosynchronous orbit, but it can lift 39,000 pounds to a space station orbit. Quadrupling the effective capacity of Earth-launched rockets will be a major improvement in space economics. Tethers would be another major advance. Right now, returning spacecraft simply throw away orbital energy when they fire their rockets to slow down. Tethers would allow the shuttle to drain off orbital speed by transferring it to the space station, where it can be used again. The only possible drawback is time: Will the savings in fuel be worth the time spent reeling spacecraft in and out on long cables?

THE INDUSTRIAL PARK

The space shuttle is now the only non-Soviet outpost for research and manufacturing in space. It provides the electrical power, the crew, and five to ten days of zero gravity. Hands-on experimentation takes place in the middeck area, where there are a number of lockers to stow supplies and carry experiments.

Middeck space is in great demand. What's available isn't for sale — researchers who want to use the middeck must be government backed or in a joint venture with NASA. 3M can run its projects on the middeck because it has joint endeavor agreements with NASA.

The shortage created an opportunity, one seized by Spacehab Inc. in Washington, D.C. Spacehab is adding an extra room to the shuttle, by creating a middeck extension to be carried in the cargo bay. It doubles the living space available on the shuttle.

The Spacehab module will occupy a third of the bay on flights where it's used, leaving room for the shuttle to launch one or more satellites. The government-run Spacelab is less flexible because it fills most of the cargo bay. Spacehab therefore may be able to win more slots on the shuttle schedule because it allows normal satellite launches. Both Spacelab and Spacehab are connected by a tunnel to the middeck's air lock.

Spacehab Inc. has collected an impressive number of customers: Tentative orders total $60 million, which would more than fill the first three flights. NASA has granted it a space services development agreement for deferred payments on six flights starting in June 1991. Spacehab won't have to come up with cash for its flights until 30 days after they return from space. For a start-up company, this financing help is crucial.

Star Wars research could be one of Spacehab's prime activities. The Strategic Defense Initiative Organization is said to have enough experiments to fill one flight by itself.

Spacehab will rent lockers with two cubic feet of space for about $1.6 million, which covers Spacehab's fee and the NASA transportation charge. Renters will get a modest level of electrical power, communications with their experiments, and some supervision by the on-board payload specialist.

The fee is more expensive than some of the alternatives, such as orbiting an experiment aboard an expendable Chinese rocket, as the INTOSPACE consortium has done. But Spacehab believes many companies will decide that the human supervision available aboard the shuttle will be worth the extra cost.

The 50 lockers riding in a Spacehab module will wipe out the current waiting line to use the shuttle middeck. The shuttle has 42 middeck lockers, but only 6 or 7 are available for science work on a typical shuttle mission. The remainder carry crew supplies, which is what they were designed to do.

Officials of Spacehab Inc. expect to raise $95 million from investors, banks, and subcontractors to achieve their goals.

The company already has signed McDonnell Douglas Astronautics to create the preliminary designs. The actual construction of the basic module will be done by Italy's Aeritalia, which has considerable experience from its work in the Spacelab program.

Spacehab's executive ranks are heavy with former top-level NASA officials. Its chairman is James Beggs, a former NASA administrator, and the executive vice president is Chet Lee, who ran NASA's office in charge of booking commercial customers on the shuttle. They gave Spacehab the skills and inside knowledge to negotiate its "fly now, pay later" agreement with NASA.

On the space operations side, Spacehab will be offering users the space-qualified hardware developed by 3M (see Chapter 7). 3M has flown experiments about a half-dozen times on the shuttle and has a NASA agreement for 62 additional experiments, making it the unquestioned expert in private space-processing work.

In addition to 3M, Spacehab also has two experienced astronauts, Owen Garriott and Byron Lichtenberg, as senior consultants.

A BIGGER SHUTTLE EXTENSION

Another private venture aims to create the summer beach house of space — a place to visit, but not permanently stay. This beach house is called the Industrial Space Facility (ISF). It's the project of a Houston company, Space Industries, that was formed in 1982. Its president is Maxime Faget, a former director of engineering at NASA's Johnson Space Center.

The ISF is to be better than the shuttle for research, but not as capable as a space station. It will be launched by the space shuttle and left in orbit. This will solve a major shortcoming of the shuttle: All equipment needed for an orbital research project must be carried into space and then wastefully brought back down when the shuttle flight is over.

It would be more efficient to stash the machinery in orbit, building up a tool chest of research equipment. The ISF will provide that permanent storage shed in space. The shuttle will visit every three or four months, and astronauts will use the equipment for several days of research. They'll also provide supplies for automatic processing units that will run while they're away. Later shuttle visits will harvest the resulting production and provide new raw materials.

The ISF will have 2,500 cubic feet, compared to the 1,000 cubic feet in the shuttle crew area or in the Spacehab unit. It will be built with two sections. A small supply module will be replaced with each shuttle visit. The unit will have a 40-inch hatch for docking with the shuttle and a 50-inch hatch for eventual attachment to the U.S. space station.

The project will cost close to $1 billion. Space Industries is trying to raise cash from private investors, and it's forming partnerships. In the fall of 1986 it won the support of Westinghouse, which will do the detailed engineering and the major construction. Outsiders estimate that Westinghouse will spend about $15 million on the project during the initial 18-month engineering and test phase.

Space Industries has won some important endorsements over the years. It attracted Joseph Allen, one of NASA's most articulate astronauts, as its executive vice president. Lockheed Missiles and Space Company, a well-regarded military space contractor, agreed to build the solar arrays. Most importantly, it negotiated a fly-now, pay-later contract with NASA. This contract, called a Space Services Development Agreement, gives Space Industries 2.5 shuttle flights — 2 for launching the ISF and half of another flight to deliver a fresh supply module. Payment for the 2.5 flights won't be due until the ISF begins generating income.

NASA has scheduled the first flight for 1992, but first, the question of who pays to construct the ISF must be settled. Space Industries had started off proclaiming the ISF as a bold private investment, but by 1988 the company was seeking a long-term NASA lease for the lion's share of the facility.

Congress initially was enthusiastic and ordered NASA to find ways it could use a "commercially developed space facility." The ISF, of course, was the only such facility being planned, but Congress wanted to appear impartial by giving it a generic name.

In fact, Space Industries wants NASA to prelease 70 percent of the ISF. Once congressional committees realized that this "commercially developed space facility" was so dependent on federal money, they began having doubts. Now the ISF faces skepticism from congressional space committees who are worried about finding the cash to pay for the space station, continued shuttle upgrades, and big space science projects.

Space Industries believes the ISF will be an intermediate step toward a U.S. space station. Companies planning to use the space station will get a chance to test equipment and processes beforehand. NASA's shuttle pilots will get practice docking with a permanent space facility. And investors in the ISF will test the marketplace for private real estate in space.

After the space station is permanently manned in 1997, the ISF could become part of its operations. The simplest role would be as a private extension to the station. The privacy could be valuable in the protection of trade secrets. The government space station will have all sorts of people aboard — NASA researchers, corporate scientists, and staffers from Japan and Europe manning their own modules. Lawyers are uncertain how to protect intellectual property—research secrets—with so many different types crowded into one space station. An ISF, as a private extension, might be able to enforce tighter physical access to corporate experiments.

It's more likely that the ISF will dock temporarily with the station to take on supplies and deliver products. Then it will move away to a companion orbit to do its work. A separate orbit gives the ISF the best zero gravity, unpolluted by the vibrations from station crew and equipment.

THE U.S. SPACE STATION

NASA began serious planning for its space station *Freedom* in 1985, expecting it to cost $8 billion and be ready by 1992. By now the cost has zoomed past $30 billion, according to the National Research Council, and a readiness date of 1997 seems more likely.

The true projected cost of *Freedom* could be as high as $80 billion to $100 billion through the year 2000, according to T. F. Rogers of the Sophron Foundation in McLean, Virginia. Rogers, a former Defense Dept. research official, directed a study of space stations for the Office of Technology Assessment of the U.S. Congress and was a founder of the External Tanks Corporation, described later in the chapter. A major driver of the higher estimate is a more accurate accounting for the cost of shuttle flights that will deliver station modules. The space agency pegged the cost of a flight at $120 million, while the actual costs run about $550 million per trip. (This results from dividing $5 billion in overall fiscal year 1990 shuttle operations costs by nine flights per year.) These higher costs, plus the likely need to purchase an additional orbiter, swell the station's costs by $15 billion, Rogers says. Other costs include inflation adjustments, depreciation costs on the shuttle fleet, and three years of shakedown costs before the station reaches full operation.

NASA officials call the space station price cheap, when spread over the ten-year construction period. Former space station chief Andrew Stofan declared in 1987 that "a few billion dollars makes no difference at all in the long term."

Others fret that NASA will design a station so costly that only a few parts will be built in this century. When faced with the bill, Congress may stretch out the construction timetable to keep spending down. Congress may even decide to kill *Freedom* at some point in its construction cycle.

A basic reason for high space station costs is NASA's founding charter.

Figure 10.1
NASA's Space Station

Source: Photograph courtesy of McDonnell Douglas Space Systems Co.

It declares that NASA must push the frontiers of technology, as it plans to do in space station computers, life-support systems, robotics, and communications. That inevitably will cost more money than simply using what's commercially available now.

Even compared to other federal agencies, NASA seems locked into extraordinarily high costs for its operational projects. T. F. Rogers notes the Commerce Dept. has placed an underwater habitat called Aquarius on the Caribbean sea floor. Aquarius will supply life support to six professionals for weeks at a time. Although they have the same volume, the space station is more complex and more remote, so it should cost more. It does — it costs *5,000 times* the less than $10 million price tag of Aquarius. Space may be 10 or 100 times more difficult to work in, but not 5,000 times.

NASA's space station is to have four major modules, two built by the United States. One U.S. module would provide sleep compartments, bathrooms, a kitchen, and an exercise area. Another would be a laboratory.

The European Space Agency (ESA) will supply a third module equipped for some type of experimental work. Congress wants to limit ESA to life sciences research. The Europeans naturally want to continue their pioneering materials-processing work done on Spacelab flights and recent Chinese vehicles. Congress fears this would let ESA skim the cream of lucrative space applications while NASA foots the bill for the overall station complex.

The Europeans almost dropped out of the space station over military issues. The Pentagon swore for years that it had no interest in the station — that way it wouldn't be stuck with any of the costs. But in 1987 the Pentagon suddenly reversed itself and asked NASA to save space for military projects. The Europeans, who sell space spending to their taxpayers strictly on the basis of peaceful exploration, were outraged.

The issue finally was papered over, with a declaration that the space station would only be used for "peaceful purposes." The Pentagon asserts that all U.S. military effort is for peaceful purposes, not aggression. By leaving the term undefined, Europe could claim victory while leaving the Pentagon free to conduct "peaceful" research.

Japan would contribute a laboratory module as well, one with an outside patio. Experiments that need exposure to space would be mounted on this platform and manipulated from within the Japanese module.

Canada is to supply a "mobile servicing center" that would roam the outside of the station. The $800 million servicing unit looks like a cross between a bulldozer and a praying mantis. It would grab approaching shuttles with a 60-foot robot arm and dock them to the station, and it would be essential to assembling the initial station modules. Later, it would tend to the experiments mounted outside the station. The servicing center would be an advanced version of the Canadian robot arm on the shuttle built by Spar Aerospace.

In return for its mobile servicing center, Canada will get 3 percent of the use of the station elements contributed by the United States, Europe, and Japan. The United States will get to use 46 percent of the European and Japanese modules as payment for all the station services those two modules will be using.

Additional small modules would be the space station's supply pantry. Figure 10.1 shows the horizontal core station modules with a shorter supply module attached vertically beneath the NASA module.

The short sections on the end of the main modules are "resource nodes" with airlocks and a clear dome on top or bottom. Crew members will stand in the dome and manipulate the Canadian robot device as it works on the outside payloads. Shuttle astronauts suggested the domes as a way to handle exterior work without requiring a space walk. Working outside in a suit is exhausting—the suits make every movement difficult and firm footing is hard to find. Getting into and out of the suits can take hours, and outside work is an inherently risky business.

Each of the eight crew members would get a compartment with 150 cubic feet of space. That's a closet five feet by five feet by six feet. The cabins would have TV and stereo, with facilities for handwashing and shaving in or "near" each compartment. Crew members would sleep in bags called sleep restraints. NASA wants the compartments to be flexible so that crew members can rearrange their walls, size, and configuration.

The station will have two toilets, two urinals, showers, and a zero-gravity clothes washer and dryer. Drinking water will be recycled from the showers and handwashers. The station might reuse water from the laundry and dishwasher, and NASA's contractors are studying whether to recycle urine.

The galley is to have a water cooler, oven, dishwasher, and trash compactor. The wardroom/dining area will seat six and serve as a crew lounge with TV and a viewing port. The sick bay is to be fairly elaborate, able to do blood tests, urinalyses, microbiology, and pulmonary tests. It's to have a digital diagnostic imaging system, such as a computerized axial tomography (CAT) scanner or ultrasound system, plus a medical data base. Crew members are to do minor surgery under local and regional anesthesia, and specifications say if that doesn't work, the station is to have "provisions for handling of a deceased crew member."

The general rules for station design call for an interior "devoid of all sharp objects." The decoration is to provide a "local vertical" in all rooms.

Figure 10.2
Diagram of Station Elements

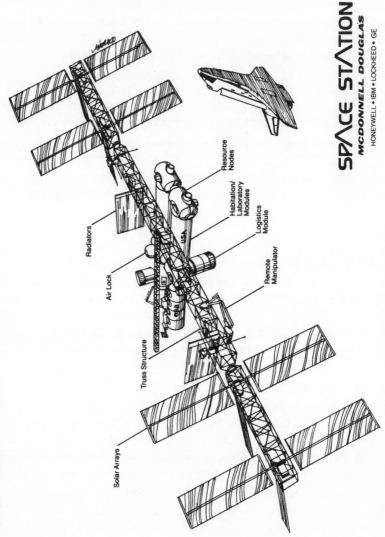

Radiators

Air Lock

Truss Structure

Solar Arrays

Resource Nodes

Habitation/ Laboratory Modules

Logistics Module

Remote Manipulator

SPACE STATION
MCDONNELL DOUGLAS
HONEYWELL • IBM • LOCKHEED • GE

Source: Illustration courtesy of McDonnell Douglas Space Systems Co.

The *Skylab* astronauts preferred rooms with an obvious up and down. When they entered a room where several walls were competing to be the "floor," the astronauts had a hard time deciding which way was up.

The station will be designed to fit almost every size of crew member. NASA standards call for equipment and passageways comfortable for "the 5th percentile Oriental female to the 95th percentile American male." These average body sizes are documented in NASA's official "Anthropometric Source Book."

Lighting will be an important design element in the station. Standards require variable-intensity light switches, or switches with high and low setting.

Allowing crew members to dim the lights, and to rearrange their living quarters, may prevent the "crotchety cosmonaut syndrome." The Soviet cosmonauts become a tad irritable after three to five months on orbit. They get touchy when ground control or fresh cosmonauts upset their established routine. That's typical of people who don't have much control over their environment, so NASA is trying to create a station that the crew can customize and control to some extent.

NASA needs to keep the crew happy to ensure that they stay productive. Wasted time — caused by fatigue or irritability — will be extremely expensive. The per-hour cost of crew time could be $17,000 or higher. With those labor rates, NASA clearly will automate everything it can. But it also wants to eliminate any distracting stress on the crew, whether it's caused by noisy equipment, harsh lighting, or lack of privacy.

Crew needs may not be obvious. On the *Skylab* mission, for example, the astronauts found their home too antiseptic. The colors were boring, and even the *smells* were boring. They became so starved for a new aroma that they gathered around to sniff the aftershave one of them had brought along.

The station will need a lot of electricity. Just the initial four main compartments and the externally mounted experiments may draw 75 kilowatts — that's enough to power seven or eight all-electric American homes. The two-phase cost-cutting plan for the station provides only 35 to 50 kilowatts, but Congress may not accept the lower level. Congress is much more enthusiastic about materials processing in space than NASA is, and has insisted that NASA provide enough electricity to power an aggressive schedule of experiments.

The station will continually expand, either with new major modules or with experiments mounted on the outside framework. Some of these outside experiments will point down for Earth studies, and others will point up to examine the stars and planets. To meet the electrical demand, NASA plans to shift from photovoltaic solar panels to solar dynamic systems in the station's second phase. These solar furnaces will focus the sun's rays into a hot beam, boiling a liquid to drive a turbine.

Solar dynamic units are more compact than equivalent solar panels, and that makes the station glide through the extreme upper atmosphere with less effort. The lower drag reduces the fuel needed to keep the station up in the proper orbit.

The latest Soviet station, the *Mir* complex, uses solar panels to generate a relatively puny 20 kilowatts.

For the shuttle program, Rockwell International was the prime contractor and organized much of the work. NASA intends to be the "prime contractor" for the station, and split up the construction of station parts among at least four major firms. Boeing will build the laboratory and housing modules, and McDonnell Douglas will build the framework that will hold the modules and the attached experiments and solar panels. The Rocketdyne Division of Rockwell will build the solar power generator, and General Electric will build satellite servicing areas and some free-flying platforms. These free-flyers will offer better microgravity than the station itself because they will be free from the vibrations caused by the station crew and its heavy machinery.

Each of the four major contractors will have almost a dozen subcontractors, who in turn will hire hundreds of their own subcontractors. For example, Lockheed Missiles and Space Company didn't win one of the four major contracts, yet it still will be paid $1 billion for various projects. Lockheed will outfit the life science laboratories and health clinic for Boeing, and it will create the station's cooling system for McDonnell Douglas, among other projects.

THE RADIATION THREAT

The space station will travel through a lethal rain of cosmic radiation. So does the shuttle, but six-day cruises don't expose its passengers to much danger. Space station workers face much higher risks from their extended stays.

Some radiation will be stopped by the station's walls. The station will have an aluminum inner pressure shell about an eighth of an inch thick that will be the primary radiation barrier. Boeing, when it designs the two U.S. modules, will try to put as much of the equipment and storage tanks around the outside walls as possible, so that these components can intercept the radiation.

Crew members are expected to be rotated every 90 days, and that's short enough to avoid dangerous radiation exposure. NASA doctors estimate the longest possible stay would be about nine months, when a crew member would hit the annual limit for radiation to the bone marrow. More on-orbit biological testing is needed because scientists don't know if they've found all the possible ways space radiation can damage the crew.

Radiation dangers drop with lower orbits because there are increasing amounts of trace atmosphere to slow down the protons crashing toward Earth. The space station will be placed relatively high at 300 miles.

NASA could lower the station deeper into the atmosphere for extra radiation protection. An orbit 250 miles high would put enough air above the station to cut the radiation exposure in half. But the thicker air would sharply increase drag on the station, slowing it down. Even the almost imperceptible atmosphere at 300 miles forces the station to fire boosters every 30 to 90 days to regain its speed. A lower orbit might require weekly rocket blasts to keep the station at the right altitude. Zero-gravity research and manufacturing would be awkward in a station doing that much hopping around.

The most peculiar part of the radiation threat is the South Atlantic Anomaly. It's a spot off the coast of South America, about 1,000 miles east of the southern tip of Brazil. The South Atlantic Anomaly is a virtual death ray of high-energy particles. The cosmic rain of protons there is five to ten times higher than anywhere else on the globe. Shuttle crews have timed their space walks to avoid being caught outside when passing through the anomaly. (In a standard orbit for launching satellites, the shuttle passes through the anomaly three times a day.)

Space station crews won't be able to come in from the cosmic rain every time they pass through the South Atlantic Anomaly. They'll need a space suit with a beefed-up radiation shield. NASA is working on several designs, including a metal one that closely resembles a medieval suit of armor.

Besides better radiation protection, the new suits will be easier to don than the shuttle suits. Astronauts must spend almost two hours getting ready for space walks using the existing suit. The reason is the shift from the normal atmosphere of the shuttle to the pure oxygen used in the suits. NASA chose pure oxygen for the suits as a way to make them more flexible. Suits full of pure oxygen can be inflated to a lower pressure than suits with a regular atmosphere — and the lower pressure makes the joints easier to flex.

The drawback with pure oxygen is the preparation time needed to start a space walk. Even before starting to suit up, astronauts must spend an hour "pre-breathing" pure oxygen to flush the nitrogen out of their bloodstreams. Otherwise, when they shift to the lower pressure of the suits, the nitrogen in their blood would start to bubble, and they would get the bends just like a scuba diver who surfaces too fast. The space station suits will be inflated to double the pressure of the old suits—using better designs to keep them flexible—and thus can use standard air instead of pure oxygen.

The new designs also aim for a better glove. The current glove is dangerous because it doesn't flex well. When astronauts need to wear it for long periods in construction projects, they have numb fingers for two weeks afterward, as the gloves pinch nerves in the fingers.

THE SPACE STATION LIFEBOAT

Finding ways to escape from the space station in an emergency suddenly became a top priority after the *Challenger* disaster. NASA officials previously assumed that catastrophic failure of their equipment simply wasn't a serious problem.

A rescue flight to the station by a shuttle might take a month or more to arrange. Keeping a shuttle continuously based at the station has been ruled out because the orbiter was not designed for long stays in orbit—and it would tie up $2 billion in equipment. Thus was born the Crew Emergency Return Vehicle (CERV).

The CERV's basic goal would be simple: to return the crew to Earth. That's something the Mercury, Gemini, and Apollo capsules did with the basic technology of a parachute drop into the ocean.

NASA's space raft is turning out to be anything but simple. In addition to being a space station lifeboat, the CERV mission is growing to include rescue of disabled shuttle orbiters as well. And it's supposed to rescue astronauts who drift away from the station during space walks. And it may be used to routinely return materials to Earth, such as purified drugs or semiconductor crystals. And it will be designed to fit on expendable rockets like the Titan IV, so that NASA could use the emergency return vehicle to *put* people into space.

Projected cost for the lifeboat is $1.7 billion, and NASA hasn't even completed the preliminary design stage. The lifeboat's escalating goals underline how NASA seems unable to keep any project limited to a single task. Instead of evolving space projects through increasingly complex stages, NASA tries to deliver the ultimate version right at the start. Consequently, it takes forever to make that start because planners are trying to guess what

Figure 10.3
The External Tank

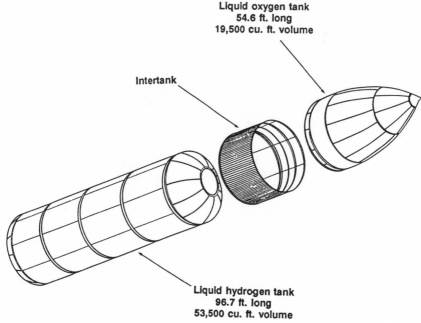

Liquid oxygen tank
54.6 ft. long
19,500 cu. ft. volume

Intertank

Liquid hydrogen tank
96.7 ft. long
53,500 cu. ft. volume

Source: NASA.

will be needed 20 or even 30 years in the future.

THE JUNKYARD SPACE STATION

The $30 billion to $80 billion NASA space station isn't the only approach to permanent space colonies. Stations could be constructed from the shuttle's external tank (ET) for much less.

The ET rides with the orbiter almost into orbit. Then the orbiter performs a special maneuver to drop it into the Indian Ocean, away from any populated areas. NASA throws away the ET as a safety precaution. It hasn't developed a method to keep it in orbit permanently—there's enough atmosphere to slow an ET down gradually. Without protection from this drag the ET would plunge to Earth after a few weeks. Burning ET chunks crashing into Tokyo, Mexico City, or Los Angeles would constitute bad public relations.

Atmospheric drag can be fought several ways. Pointing the ET nose first into its orbital path would create less drag than if it were flying sideways or tumbling. Martin Marietta, the tank's builder, estimates a nose-first orientation would keep an ET flying for about a month at 160 miles altitude. For more permanence, the ET must be boosted to a higher orbit where the atmosphere is thinner. Whatever method is used to get them higher, storing the ETs in orbit creates a new resource for space commerce. The tanks could be the shell for low-cost private space stations, and others could be taken apart to create an extremely valuable junkyard, supplying material for budget-conscious space pioneers.

If the shuttle took the ET along into orbit, it could fly a direct-injection trajectory that is more efficient than the shallow path usually flown. NASA already has tested the direct-injection trajectory, so the idea of taking ETs all the way into orbit seems sound.

Taking the ET into space also will provide extra fuel for use in low Earth orbit. The fuel is the residue left over in the ET's two sections. Figure 10.3 shows how the ET is constructed. It's actually two tanks joined with an "intertank" structure.

About 15,000 pounds of liquid hydrogen and oxygen remain in the two tanks after the ET reaches orbit. They can be scavenged to provide propellant for the space station, for orbital transfer vehicles, or for missions to the Moon and Mars. The oxygen and hydrogen also could be combined in a fuel cell to generate electricity. The fuel cell's waste product is pure water, which space pioneers would otherwise have to import from Earth.

The ET is composed of 53,000 pounds of aluminum, and this material could be used for fuel as well. When turned into powder, aluminum burns quite fiercely in combination with oxygen.

The aluminum in the ET is aerospace-grade strength, already formed into factory-tested pressure-tight shells. The hydrogen tank in particular could become a low-cost space habitat. The tank is something like a "handyman's special"—it needs some fix-up work but can become a home for a dozen or more space workers.

The hydrogen tank is 27.5 feet wide—wider than the average home's living room—and 96 feet long. The home's front door may be an inspection manhole that is already part of the design. The manhole is three feet wide, big enough for a space-suited astronaut to enter. To make a custom home, the dome that caps the tank can be unbolted to completely open one end. The dome could be hinged to create a huge hanger door, or it could be replaced with some specialized apparatus.

The hydrogen tank has tremendous volume. Its 53,500 cubic feet is eight times the area proposed for each space station module and double the pressured volume of the entire station. That wide open space has excited a lot of imaginations. A report prepared for the Space Studies Institute says a fully functional ET space station could be launched on a single shuttle flight. The cost would be only $200 million to $300 million, according to the report. Inflatable room dividers would quickly split the tank into work spaces, living quarters, and a kitchen/dining area.

The Space Studies Institute report advocates an addition to the ET called an aft cargo carrier (ACC). In the "instant space station" plan, the aft cargo carrier would house the high-tech equipment needed for the station. It would carry the communications gear, the computers, and the life-support systems. An air lock for space walks also would be built into the ACC. All the complicated parts of this instant space station would be constructed on the ground as part of the ACC and the hydrogen tank with inflatable partitions would take the relatively simple role of crew quarters and workshops.

In normal shuttle launches, the ACC would carry cargo that couldn't fit in the payload bay. (The shuttle often runs out of volume in the payload bay before it runs out of lifting capacity. Payloads have turned out to be bulky rather than heavy.)

Some additional high-tech equipment, such as a back-up life support system, could be carried in the payload bay. It would be added to the ET after it reaches orbit. The result is an affordable space station. Several could be lofted, to serve corporations and university research consortiums. A space station completely owned by 3M, for example, would not have any trouble protecting 3M's trade secrets. By comparison, NASA's all-in-one multinational space station faces problems just sorting out which country's laws would be effective in which modules. Would a Japanese experimenter working in a module paid for by Japan and partially built in Japan potentially be subject to U.S. patent rules? If a French scientist steals the idea of a German scientist on the station and radios down to get a European patent, will the resulting fight be decided by U.S. standards or European rules? Avoiding all these questions with a private ET space station will appeal to many corporations.

THE EXTERNAL TANKS CORPORATION

A start-up venture called External Tanks Corporation in Boulder, Colorado, is eager to start building these instant space laboratories. The company is in partnership with the University Corporation for Atmospheric

Figure 10.4
Labitat Made from External Tanks

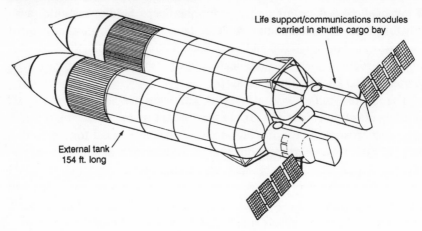

Life support/communications modules
carried in shuttle cargo bay

External tank
154 ft. long

Source: Illustration courtesy of External Tanks Corp.

Research, a consortium of 57 institutions that define atmosphere as every-thing out to the rim of the solar system. A foundation created by the con-sortium owns 80 percent of External Tanks.

The remainder is being sold to potential corporate customers and to in-vestors like Thomas Rogers, one of the company's founders. Rogers was a deputy director of research and engineering at the Pentagon and is adamant that government projects will *never* lead to a space-faring civilization. Rogers directed a study of civilian space stations for Congress's Office of Technology Assessment and became convinced that ETs offer a way to put private enterprise in charge of some potentially valuable space assets.

The company thinks it can convert two ETs into a Spartan and unfur-nished space laboratory for $100 million (see Figure 10.4). The firm's "Labitat" facility would cost $15 million a year to run, including two crew members responsible for maintenance and repairs. Corporate users would bring their own researchers, equipment, oxygen, water, and food.

External Tanks expects annual operating costs to run close to $200 per cubic foot. The firm expects to rent space to commercial customers for $565 per cubic foot and to nonprofits for $285. The company predicts gross profits of $26 million to $32 million in 1992/96.

The company won important start-up funding in July 1987 from Autodesk Inc., which markets engineering software for personal computers. Autodesk paid $225,000 for 5 percent of the firm's stock. Autodesk president Alvar Green said the private space station would be a demonstration project — engineers would use Autodesk software to prove the value of PC-based design tools.

The Labitat is similar to the ISF planned by Space Industries. Both would be permanent space outposts available for corporate and scientific research. Labitat would be 50 times larger, and continuously staffed instead of periodically visited like the ISF.

GLOBAL OUTPOST

External tanks also can be used for a space station's other major role — transfer point and way station. Global Outpost Inc., of Alexandria, Virginia, is readying plans for an unmanned external tank to store fuel, water, and power.

The Outpost ET would include tether services, such as the downward release of tethered satellites to allow their reentry without expending valuable fuel for deceleration. The platform also would serve as a testbed for advancing tether technology.

Tethers can't be used on a space station doing microgravity work, such as materials processing. Capturing and releasing satellites would cause too much vibration for material processing, and tethers also distort a station's center of gravity. Zero gravity on a space station is never absolutely zero, but comes closest at the exact center of the mass, where centrifugal force exactly balances the pull of gravity. Moving even a few feet higher or lower than the center brings measurable pull on an object. Tethers send part of a station's mass very far away, shifting the center of gravity. With a heavy satellite on a 200-mile tether, the center of gravity may not even be inside the station anymore.

The Global Outpost project seeks to be the first, and only, space structure with the freedom to exploit tethers and other jobs with a lot of pushing and shoving, like fuel and water transfers.

The enthusiasm for private space stations is remarkable. Four major efforts are under way despite the *Challenger* disaster. Spacehab Inc. is attracting seasoned executives for its extension to the orbiter cabin; Space Industries has snagged Westinghouse as a major partner for the ISF; External Tanks has the backing of a substantial academic foundation to build private space habitats with revolutionary low costs; and Global Outpost is staking a claim to the role of orbital way stations for tethers and refueling.

PRIVATE PROJECTS MAKE SPACE AFFORDABLE

NASA's space station will be 100 to 200 times more expensive than the External Tanks or Global Outpost station, and have only half the volume. That's a 100:1 advantage for private enterprise over governmental efforts if these two firms can meet their cost projections. True, the ET versions are bare bones while NASA's station will come fully furnished with laboratory equipment, new space suits, and scores of experiments bolted to the outside framework. So perhaps the actual advantage is only 50:1 or 20:1 in favor of private space projects. That's remarkably similar to the edge Geostar holds over the military in creating a navigation system. Free enterprise can slash the cost of space projects, if given a chance.

With only modest cooperation from NASA and some initial customers, these enterprises could grow into a real space services industry. Along with the private booster companies, they will finally blaze a path to space development that's cheap enough to afford.

In the late 1990s, the space economy will begin to flourish. Private space stations using shuttle external tanks will fuel a real estate boom. Then spaceplanes will cut launch costs, and demand for space transportation will surge. The combined real estate and transportation booms will make many companies and investors rich.

The economic center of the space trade will be low Earth orbit, 250 to 400 miles up. At the turn of the century, hundreds of people in dozens of outposts will be doing research, making electronic crystals, refining drugs, and building new space stations. Others will be in service jobs, repairing satellites, refueling orbital transfer vehicles (OTVs), or running their station as an energy bank with tethers capturing and releasing payloads.

Some years later, orbital construction crews will arrive for two major projects: (1) the first solar power satellite to beam electricity down to Earth, and (2) a ship for the first Mars expedition. Both projects will spur increased demand for transportation between the ground, low Earth orbit and the Moon. The solar power satellite almost certainly will use construction materials quarried from the Moon, and the Mars expedition may need Moon oxygen for fuel.

No one can predict when this phase will begin—at the turn of the century, in 2010, or much later. New technologies often arrive much faster than anyone expects. A *Popular Mechanics* article in March 1949 noted that while the ENIAC computer then weighed 30 tons, "computers in the future may. . . perhaps weigh only one and one-half tons." Sophisticated pocket calculators actually weigh a few ounces now, of course.

Earlier, in 1912, the federal government actually indicted inventor Lee DeForest for claiming his audion tube would allow cross-Atlantic radio conversations. A U.S. district attorney prosecuted DeForest for mail fraud because the inventor said "that it would be possible to transmit the human voice across the Atlantic before many years. Based on the absurd and deliberately misleading statements, the misguided public has been persuaded to purchase stock in his company."

DeForest was later called the "father of radio" because the audion tube was crucial in radio, radar, and television until the invention of the transistor. Yet the government attack on DeForest came *after* he'd been granted a patent on the audion tube, and *after* cross-Atlantic telegraphy had been demonstrated by Guglielmo Marconi in 1901. DeForest made the mistake of having advanced ideas of his own that weren't shared by federal prosecutors and other experts of the day. Any member of society's flock who strays too far from the "common wisdom" risks getting his eyes pecked out.

Still earlier, the *Boston Post* in 1865 applauded the arrest of Joshua Coopersmith for daring to raise funds for telephone research: "Well-informed people know it is impossible to transmit the voice over wires and that were it possible to do so, the thing would be of no practical value."

Even after Alexander Graham Bell succeeded in his telephone work, the experts were not impressed. President Rutherford B. Hayes tried an experimental Washington-to-Philadelphia phone call but couldn't imagine who'd want to use such a thing. The Western Union Telegraph Company, staffed with the leading communications experts of the day, agreed with Hayes. When Bell offered the telephone patent to Western Union for $100,000, the company rejected the deal. The experts couldn't envision what use it would have.

An imagination barrier plagued these inventions, slowing their adoption and blinding people to their implications. Space development suffers the same barrier. As a result, even projects with obvious merit must beg for money. The space economy will bloom only when the imagination barrier crumbles, at some unpredictable future point. That's why the dates chosen here for various advances are arbitrary—they depend more on mass psychology and investment decisions than they do on technology.

ELECTRIC UTILITY PLANTS IN ORBIT

Local utilities trying to build power plants increasingly hit the NIMBY problem—"not in my back yard." Residents near the proposed site argue that a utility plant should be built "somewhere else." They don't want the soot from coal or oil, and they certainly don't want a nuclear plant nearby.

A utility's struggle for a site is followed by a struggle to keep up with constantly tighter pollution standards, complicated by erratic swings in fuel prices. A utility must decide to run this gauntlet long before it actually needs the power, in some cases ten years in advance of the predicted need.

Utilities could solve these problems with zero-pollution solar power satellites. Solar power is still a marginal technology on the ground, but in space it really makes sense. Space solar arrays don't cover up valuable real estate, don't get blanked by clouds, and don't quit when the sun goes down. A solar power satellite would spread out its panels over perhaps 50 square kilometers and then beam the power down to Earth via microwave or laser. A very low-tech "rectenna" on the ground would feed the power into a utility's grid.

A solar power satellite also solves the utility's problem of predicting electricity demand a decade in the future. A utility that starts to build a solar power satellite and discovers ten years later that it's not needed can sell the excess capacity anywhere in the world. The beam is simply directed to a paying customer somewhere else until the home utility needs it.

Excess power is sold to other utilities now, but only on a regional scale, across a few states at most. Too much energy is lost to transmission line leakage for long-distance exports. Solar power satellites don't have such limits and will give utilities the entire globe as a marketing region.

The only full-scale study of solar power satellites came in 1979 during the hysteria of the oil crisis. The Energy Department and NASA spent about $30 million for reports that concluded solar power satellites were too expensive to compete against terrestrial fuels. Just boosting all the materials into geosynchronous orbit might cost $500 billion for a five gigawatt satellite.

The $500 billion would buy a lot of electricity from space — five gigawatts is more power than any Earth plant generates. All the plants in the world together add up to only 2,000 gigawatts. Most studies focused on huge satellites as a way to get economies of scale to make the electricity affordable.

In 1985, a new private study showed how a solar power satellite could be constructed for much less. The Space Studies Institute calculated that lunar materials could be used for virtually the entire project. Only 500 tons would be lifted from Earth. That would cost only $500 million to launch, assuming mass-produced boosters bring launch prices down to $500 per pound.

The $500 million for Earth components (things like switches and microwave tubes) is a lot of money. To that, add about $1 billion for shipping 50,000 tons of lunar supplies to geosynchronous Earth orbit. The $1 billion for Moon materials is based on an estimate that they will have a 50:1 advantage over Earth supplies in transportation costs to geosynchronous orbit. The advantage is due to the weak lunar gravity, which requires much less energy to overcome than Earth gravity.

Moon materials have the added advantage of not requiring expensive rocket propulsion. Electric mass drivers can hurl supplies from the Moon to geosynchronous orbit, unlike Earth materials that must be lifted by costly chemical rockets. Combined Earth and Moon launch costs may hit a very substantial $1.5 billion. Mining on the Moon and assembly of the station in orbit would be added expenses.

But the gigantic costs will be balanced by huge benefits. A five gigawatt power station would produce electricity worth *$2.5 billion a year*—year after year with no costs for buying coal, oil, or uranium. Only maintenance costs for the station and the ground receivers would detract from the profits. Even if Moon mining and orbital assembly are double the transportation costs, utilities still could start making profits after only a few years of operation.

Creating the first solar power satellite will take some courage in testing new technology. The risk can be shared by creating a satellite to serve peak power needs and splitting the ownership across several continents. Split ownership makes sense because electrical demand is uneven. Demand peaks during the day, usually in late afternoon. A solar power satellite could shift its beam and serve that peak demand repeatedly.

One scheme would have the satellite do eight-hour shifts at each of three cities around the globe. It could start beaming power down to Tokyo at 11 A.M. Eight hours later, after factories and offices have closed and evening meals are cooked, the beam could shift to Europe. In Rome, for example, local time would again be 11 A.M. After a day's work in Rome, at 7 P.M. the beam would shift once more, perhaps to Denver or Phoenix where the local time again is 11 A.M.

Electricity for peak demand always is the most expensive to generate. A peak power plant stands idle most of the day, and that's costly. A solar power satellite therefore could replace the most expensive power that utilities now produce. That would make construction of this new technology easier to justify, and the financial risk of building the first solar power satellite would be less when split three ways in a peak-power scheme.

To reach three cities spaced evenly around the globe, the solar power station could be placed in a very high orbit with a 48-hour period. With the Earth turning every 24 hours below, the effect is to create a sun-synchronous satellite—it appears overhead at the same time every day and delivers power for the eight hours of highest demand. Another solution would be a sunsat in geosynchronous orbit over one customer, with relay stations above other customers using reflectors to bounce the power down to them.

WORLD WARD III SURPLUS

Utilities aren't the only customers for gigawatts of electrical power. The Strategic Defense Initiative, if deployed, will need huge bursts of energy for its beam weapons. The Pentagon is researching all sorts of orbital power supplies, from nuclear reactors to batteries and giant flywheels. Many of these would be more compact and easier to protect than a solar power satellite spread over 50 square kilometers. But with ample lunar material for shielding against laser damage, a solar power satellite could be quite "survivable," as the military likes to say. It's *so* spread out that burning a hole through hundreds of square meters of solar panel simply won't have much effect. The core unit with the electronics would be the only part needing heavy shielding.

Actually, laser weapons aren't expected to actually burn holes in space battles, because the distances are so great that the beams won't be tightly focussed. The beams will heat their targets, but not actually burn through them. For this reason, a solar power satellite is harder to destroy than a rising nuclear missile because it is not under dynamic stress. A satellite's laser beam can take out a missile simply by heating it unevenly, causing metal to expand on one side, making it explode or go out of control because of the stresses of the launch. A solar power satellite, by contrast, is at rest in space. A ground-based laser can heat one section without causing it to tear apart violently.

If solar power satellites meet Pentagon criteria for beam weapons, military mass production would make them cheaper for utilities. Costs would be even lower if the same satellites could be used for war *and* peace. A solar power satellite could serve both masters. It could keep our toasters running until a crisis, and then it would start toasting Soviet ICBMs. (Or Libyan ICBMs, if Muammar Qaddafi buys rockets to carry the atomic weapons he already has.) Solar power satellites could be World War III surplus, available before the war instead of after it.

If SDI doesn't use solar power satellites, it still will need plenty of shielding material. The Moon is the logical place to get it. The military need for Moon material will bring economies of scale for mining and refining operations there, indirectly subsidizing the costs for solar power satellites.

Cheap electricity from space is inevitable. The only debate is over how quickly transportation costs will drop to make it feasible. Perhaps it will take 15 years before construction starts, or perhaps it will take 30 years. At some point, it will make sense — and then the new Moon age will begin as companies build mining camps and smelters.

RETURN TO THE MOON

Remember the Apollo lunar landings? Astronauts did kangaroo hops and zipped around in a Moon buggy. It was a great tourist trip until they got to the souvenirs. All the astronauts brought home were gray rocks. And their home movies were just as bad—a worthless sea of gray dust, sprinkled with gray boulders, with gray mountains in the distance. The place was boring and few taxpayers saw the need to continue Moon visits.

Chemical analysis of lunar soil shows it's much more valuable than it seemed at first glance. More than 40 percent of lunar soil is oxygen, the main propellant in oxygen-hydrogen rocket engines. These engines use six pounds of oxygen for every pound of hydrogen, so producing lunar oxygen would slash the amount of propellant that must be boosted from Earth to refuel satellites and spaceships.

About 20 percent of lunar soil is silicon, the main ingredient in glass. Solar power satellites will need glass covers and backs for the power cells, which themselves can be made from silicon. Glass also can be the main ingredient in composite building materials. Composites, which have more strength per pound than most metals, were used in the round-the-world *Voyager* endurance test.

The two most abundant minerals on the Moon exactly match the two most pressing needs of space development: cheap fuel and cheap construction materials.

The lunar oxygen is locked up in several types of metal oxides, including silicon, aluminum, titanium, and magnesium. Releasing the oxygen will produce valuable metals as by-products. Wolfgang Steurer of the Jet Propulsion Laboratory calculates that simple solar heaters can cook lunar soil at 3,000 degrees Kelvin to produce oxygen and the various metals.

A factory producing 100 tons of oxygen per year could be built on the Moon with only 14 tons of equipment carried from Earth. With a ten-year life and 10 percent of the product lost through spills and accidents, the plant could produce 900 tons of oxygen. That's a payoff of 63 tons of fuel for every ton of factory carried to the Moon, according to Steurer's estimates. Lunar miners can sell that oxygen on the Earth-Moon-Mars trade route for far less than oxygen brought up from Earth.

Spaceships working the trade route will still need hydrogen, which hasn't been found in the Moon's table of contents. They may pick it up in Mars orbit from water cooked out of the soil of Phobos, a Martian moon. Until a Mars mission is mounted, however, space traders will look to the Moon for hydrogen substitutes.

Powdered metals might be one substitute for hydrogen. Powdered aluminum or magnesium burns with less thrust than hydrogen, but these metals would be vastly cheaper for in-space work than hydrogen lifted from Earth.

Steurer's hypothetical Moon factory doesn't include any machinery for recovering the metals produced. Adding that equipment would increase the 14-ton shipping mass by an unknown amount, and yield an array of metals. The metals aren't mere trace elements. Depending on the site, Apollo astronauts collected soil that ranged from 5 to 14 percent aluminum, 4 to 15 percent iron, and 0.29 to 5 percent titanium.

Most of the metals are combined with oxygen and require processing to release the pure metal. But a lot of pure iron is freely available as tiny blobs sprinkled throughout the lunar soil. The blobs are the residue from meteors that have bombarded the Moon for eons. A robot tractor dragging a magnet through the soil would easily gather these iron bits. The tractor could collect iron equal to its own mass every hour, according to Robert Waldron of Rockwell International and David Criswell of the California Space Institute.

Lunar Power Stations

Waldron and Criswell propose that a robot tractor could do several jobs at once. In addition to collecting iron, it could build solar power arrays as it plows the lunar soil.

The tractor would create long, small ridges to catch the sunlight. Then it would spray layers of glass, iron, and silicon on the ridges to create the solar collectors. The electrical circuits can lie right on the surface because the lunar soil is bone dry. No moisture means that no electricity bleeds away to the soil, and the parts don't corrode.

Criswell and Waldron propose two solar farms on opposite edges of the Moon so that one produces power while the other sits idle through the 14-day lunar night. Despite working only half time, the lunar power stations could be very profitable because their construction costs are so low. They would be built with lunar materials, just as a solar power satellite could be, but the lunar materials wouldn't have to be launched back to geosynchronous orbit. Construction on the Moon is vastly simplified; no supporting structure is needed and the robot tractors don't have to be very sophisticated to do the job.

Criswell and Waldron say the electricity can be microwaved back to Earth with more accuracy than a solar power satellite. Earth sites could receive Moon power for eight hours a day. For continuous power, microwave reflectors in orbit could keep the beam steady on a single site all day. Criswell and Waldron estimate that ground rectennas to receive the lunar power would cover only a tenth as much area as an equivalent solar photovoltaic system.

The Lunar Cement Plant

The Moon will attract real estate developers because the workers creating oxygen and power for export will need someplace to live. Imported housing will be so expensive that few could afford it. Building with local materials is the only economical way to create lunar homes and industrial parks.

Developers could mine lunar metals to create iron or aluminum sheets for construction. That would require a boat load of heavy equipment from Earth—mining machines, a refinery and smelter, and some type of metal-forming equipment.

A much simpler construction approach for the first homes and offices is the "lunarcrete" method. All that's needed are molds in which to pour ordinary Moon soil, and some heaters and vibrators.

Researcher Larry Beyer of the University of Pittsburgh says a lunarcrete project would begin with the shipment of two molds and a work crew from Earth. First, the workers would bolt together an inner mold. It would be pressure tight and would serve as their temporary quarters. Then they'd erect an outer mold with a six-foot gap from the inner mold.

The crew would pour lunar soil in between the two forms. Vibrators attached to the outer form would constantly shake the soil into denser and denser compaction. In nine days the crew could fill and compact a form enclosing 2,300 square feet, about the area of a suburban house.

Then the lunar settlers would remove the outer forms and set up mirrors or microwave heaters. The energy beamed against the outside wall would glaze it, melting the surface into a glass-ceramic layer that's completely airtight. Beyer estimates that the equipment needed—forms and heaters—might weigh 37 tons and require two shuttle flights to lift into low Earth orbit.

The cost of shipping the forms and crew from Earth would be a one-time expense. After the first house is complete, the forms can be used repeatedly to create an entire tract of lunarcrete homes and factories. The initial crew would keep making "shake-and-bake" shells without any further supplies from Earth.

Beyer's process is adapted from a standard terrestrial method called the dry, vibratable concrete process. It's used to line induction furnaces, blast furnace troughs, and steel mill ladles.

Even the furnishings inside the shells can be made from lunar materials. Gordon Woodcock of Boeing Aerospace predicts that about 90 percent of a lunar base could be built directly from lunar materials. Things like inside walls and shelves could be made from sheet metal (aluminum and iron) or glass-glass composites coated with a solar-fired glaze for smoothness. Plumbing and small air ducts would be manufactured from lunar metals. Large low-pressure air ducts could use glass-glass composites or even basalt rock that's been melted and cast into tubes. Gears, housings, and other machine parts could use powder metallurgy—powdered metal is poured into forms, heated, and cooled to form strong pieces. By using molds, powder metallurgy can shape metal without the huge metal stamping machines found in factories.

The first lunar factories will be devoted to expanding and furnishing the colony, according to Woodcock. He believes that producing oxygen for export will be secondary. Cutting back on Earth imports as quickly as possible has a greater economic payoff.

Woodcock's studies show that building with local materials and extracting lunar oxygen makes a huge lunar base fairly affordable. The major cost is getting the initial equipment and crew on the surface. Woodcock compared annual operating costs for a six-person base with a 1,000-person base. Costs for the huge colony would be less than double those for the tiny outpost if lunar materials were put to work.

Woodcock only estimated the relative costs of small versus large bases. The actual cost of an initial six-person base, for example, depends on when it is created. The later the start-up, the less expensive launch costs to reach the Moon will be. Even with today's prices, a lunar base will cost much less than the Apollo visits did. Woodcock estimates that one man/day of an Apollo Moon visit cost the taxpayers $160 million in 1986 dollars. One man/day at a six-person lunar base may cost only $10 million. The cost of a 20-person base drops to $1 million per man/day, assuming all supplies come from Earth—no lunar materials for the first crews.

That puts the burden of a 20-person base at over $7 billion a year. Congress isn't likely to pay for it. But consider the double impact of the next generation of cheaper rockets and the savings from building with lunar materials. Launch costs by the late 1990s likely will be only a fourth of what they are now. And a Moon base using lunar oxygen and building supplies costs only a tenth as much as an all-Earth version.

Together, these forces will push the annual spending needed for a substantial lunar base lower every year. At some point in the steady decline of costs, Congress or corporations will decide the expense is justified by the potential gains.

Science and Wealth from a Lunar Colony

In addition to the Moon's economic value, it also offers science payoffs. Wendell Mendel, a leading NASA Moon expert at the Johnson Space Center, lists five major science goals for a lunar base:

Learning the origin and history of the Moon. We don't know how it was formed or whether it has an iron core like Earth.

Astronomical observatories. Telescopes built on the Moon can have 10,000 times the resolution of the Hubble Space Telescope, and 100,000 times the resolution of Earth instruments. The sharper images would let scientists inspect other stars for solar systems.

The far side of the Moon permanently faces away from Earth and is an ideal site for radio telescopes. It would be a quiet zone, shielded from the babble of radio signals that blankets the Earth.

Physics and chemistry. The Moon offers a high vacuum. In the permanently shaded polar regions, it also provides a deep-freeze approaching absolute zero. Physicists studying general relativity could build a good observatory for gravity waves.

Origin of life. A complete exploration of the Moon could provide clues on how life began. Meteorites that hit the Moon billions of years ago may still be in their original state, since the Moon doesn't appear to have any water. They may show if Earth was "seeded" with the building blocks of life by meteor showers.

Learning to live in space. Creating a self-sufficient lunar base will teach us how to live in space. Mistakes and oversights can be corrected while conveniently close to home. The lessons learned—from life-support systems to crew selection—will be invaluable when a Mars base is attempted. Miscalculations on a Mars mission would be harder to correct—emergency supplies to fix a problem could take two years to arrive.

The science missions logically should be financed by the taxpayers. NASA would manage the science operations, but there's no reason why it also should create a second Apollo program and supply every piece of needed equipment. Congress could direct NASA to buy its transportation and housing from private sources. This would give a great boost to corporate Moon development. The initial NASA contracts would give firms a foothold in exploiting the Moon's economic resources. Companies serving government science bases could branch out into such things as low-cost lunar oxygen to fuel the Earth-LEO-Moon-Mars trade routes and lunar metals and composites to slash the cost of space stations and solar power satellites.

Even more daring would be a completely private expedition, an idea gaining more credibility as researchers study the possibilities. Several investigators are proposing a lunar base made from part of the shuttle's external tank (ET). As noted in the previous chapter, the ET has three main parts — the large hydrogen tank, a smaller oxygen tank, and a skirt that bonds the two togerther called the intertank. The oxygen tank and skirt can be detached from the ET and form the core of a low-cost lunar base.

One plan presented by Charles King of the Bionetics Company in Hampton, Virginia, would create a 12-person lunar habitat by remodeling the oxygen tank and intertank in low Earth orbit. The oxygen tank would be outfitted with flooring, life support and power equipment and an airlock. RL-10 rocket engines would be bolted on to the intertank. After conversion, it would be sent unmanned to the Moon.

The conversion equipment would weigh only 60,000 pounds, a little more than one shuttle load. The propellant amounts to 178,000 pounds, and probably would be lofted on unmanned rockets for safety reasons.

After the base arrives safely, the crew would come over in a lunar lander. Delivering the base pre-built helps make the lunar lander more economical. The lander can be designed in a compact size for its long-term role of crew resupply — it needn't be huge (and costly) for delivery of construction materials for the base.

Remodeling the ET in orbit as a lunar base will require substantial work outside — extra-vehicular activity (EVA) in NASA jargon. The space agency, as noted in the previous chapter, is trying to create space suits flexible enough — especially in the gloves — for long-term EVA. NASA is spending millions on the effort and the resulting suits will be so expensive that the agency isn't planning to make enough sizes to fit all the potential space station crew. (No suits are planned for women who are smaller than the median female body, for example.) One private firm hopes to overcome this cost barrier with an entirely new approach.

Cis-Lunar Development Laboratories, using private funds, is building a prototype space suit based on sea diving technology. The suit will be inflated to very low pressure, about three pounds per square inch (psi), compared to the eight to fourteen psi planed for the NASA suits.

President William Stone notes that low pressure means the suit will be more flexibile than the tightly inflated NASA versions. Low pressure also should be used inside crewed spacecraft and lunar bases, Stone believes. Because they would be holding back less pressure, spacecraft and bases could be built with greater safety margins, making the expensive process of design and testing shorter.

The Cis-Lunar spacesuit may cost only $100,000, which is less than 5 percent of the NASA suit's cost.

With innovative ideas like the ET lunar base and truly affordable space suits, a private Moon expedition would cost a small fraction of Apollo's price tag. The explorers could pay their way by selling the media rights, and by bartering with corporations for equipment.

For example, NBC paid $401 million for the U.S. rights to the 1992 summer Olympics. CBS paid $1.1 billion for four years of World Series and playoff games. A daring return to the Moon by private adventurers would be at least as valuable as one summer Olympics, and may approach the value of a few years of World Series. The TV rights in Europe and Asia would add to the cash available. Global broadcast income for a Moon project could total from $1 billion to $2 billion.

The expedition would gain additional support from corporate contributions and bartering. James Dunstan, a partner in the Washington law firm of Haley, Bader and Potts, notes that an expedition could barter with companies for computers, medical equipment, communications devices and the like. The expedition would get free supplies and the corporations would get valuable publicity rights — "The Official Personal Computer of the Moon Expedition," for example.

The 1992 summer Olympics is expected to rack up about $400 million in such corporate sponsorships. Coca-Cola paid $30 million to be a sponsor, in addition to the advertising time it will buy on NBC.

With at least $2 billion available from the media and corporations, a private lunar expedition would have all the cash needed for a successful break with history.

Whether the first permanent base is private or NASA-funded, it will set the stage for the next big adventure—the trek to Mars.

THE MARS EXPEDITION

The first travelers to Mars probably will start their journey in 15 to 20 years. Declining costs almost ensure the first expedition will leave in the first decade or two of the twenty-first century. Second-generation spaceplanes and lunar oxygen will force costs down gradually each year. Eventually, a government or a private consortium will grab for the glory.

The first Mars expedition most likely will aim for its moons, Phobos and Diemos. They are easy targets every two years when Mars and Earth are closest. At these times, the Martian moons require less fuel to visit than our own Moon. That seems impossible because they are so much further away. But in space, distance is less important than gravity. Fighting the gravity of a planet or moon is the main challenge.

The Martian moons are perfect destinations—they are so small that they have almost no gravity at all. Ships will rendezvous with Phobos as if they were docking with a space station. Returning to Earth requires only a modest shove, compared to the launch from our Moon's gravity well. The low-thrust return from Phobos every two years is the reason that total energy required for a visit can be less than for our Moon.

The first expedition likely will set up camp on Phobos and drop robot explorers down to Mars. The robots would be cheaper to land on Mars than a crew because they don't eat, breathe or expect to return home. The robots would be actively guided by human controllers on Phobos. Active control isn't feasible from Earth because of the 20- to 40-minute delay in sending signals round trip. A robot explorer could topple off a cliff before its boss back at the Jet Propulsion Lab in California got the first image of the approaching danger.

The crew on Phobos also may set up a factory to produce carbon dioxide and water from the moon's soil. A simple unit the size of an office desk would focus solar energy to heat rock to a few hundred degrees Celsius. The water then would be split into hydrogen and oxygen for rocket fuel. The hydrogen may be tough to store because it requires supercold temperatures to stay liquid; The crew may convert it to methane through a catalytic reaction with the carbon dioxide.

The fuel churned out by the Phobos factory would be "free" in that the expedition wouldn't have to cart it all the way from Earth. The initial Phobos fuel could be earmarked to provide the extra energy needed for a manned Mars landing from the Phobos base. If the factory didn't work out, the first expedition would remain on Phobos and explore Mars via their robot teams.

NASA's Office of Exploration sketches a more modest Phobos expedition that arrives in May 2003. The four-person crew would explore the moon and control robot rovers on Mars, but wouldn't attempt to land on Mars themselves.

The NASA mission uses a "split-sprint" tactic. The mission is split into cargo and crew spacecraft. The cargo is launched first, in February 2001, on a slow, minimum-energy trajectory. The cargo ship carries the Phobos exploration equipment and the propellent needed by the crew to return to Earth. It would arrive in Mars orbit in October 2001.

The crew would blast off from orbit in August 2002 on the "sprint" part of the mission. In November, they'd swing past Venus to shape their orbit for the best approach to Mars. In May 2003 the crew would arrive in Mars orbit. Two of them would depart to explore Phobos and two would remain on board to teleoperate rovers exploring the surface of Mars.

The team departs 20 days later, passing by Earth in October 2003. The crew would abandon their ship and land with an Apollo-style atmospheric reentry vehicle.

A Phobos mission arriving in 2003 takes advantage of the best possible alignment of planets for a "sprint" mission, for the lowest possible fuel requirements. NASA calculates the 2003 target is an opportunity that comes around once every 32 years.

The 2003 mission would require that 1,779 tons of equipment and fuel be placed into low Earth orbit for the mission. Waiting until 2005 swells the required mass in LEO to 2,365 tons.

The chief flaw of NASA's Phobos plan is the loss of the Earth-Mars spaceship, which the crew abandons upon reaching Earth. The plan ought to result in the creation of a permanent Earth-Mars vehicle that could be used in the next step, the actual landing on Mars. At the very least, the Earth-Mars ship should be recovered for examination by engineers designing the next vehicle.

On the positive side, a quick Phobos visit could set up a small propellant factory, proving its feasibility for later Mars landing missions. At this point, we have no guarantee that Phobos is the right composition to allow production of oxygen and hydrogen fuel.

The Soviet Union hoped to get some definite answers early in 1989 with a pair of unmanned Phobos probes. One died on the way, when it was accidentally, and permanently, turned off. A technician ignored procedures and sent incorrect computer instructions.

The remaining probe carried a laser to zap Phobos as it flew by. Equipment was going to sample the ions kicked up by the laser blast. The analysis of the ions would have revealed the chemical makeup of Phobos. The remaining probe managed to reach orbit around Mars and take limited photos of Phobos. But Soviet controlers lost contact with the probe before it could examine Phobos with its laser or radar.

Both probes were lost during maneuvers when the craft were taking photos. To save money, the Phobos probes did not carry a platform that could scan objects independently of the main craft. Instead, the entire probe had to be slowly turned to complete a scan. This interrupted normal radio contact as the antennas pointed away from Earth during the scan. Pinching pennies by eliminating a scanning platform was a key error. So was the lack of a fail-safe mode, in which the spacecraft would automatically fall back to a stable waiting position if contact with its controllers was interrupted.

The Soviets had expected to follow this crucial experiment with a 1994 mission to Mars itself. It would land a rover on the surface to take measurements and photos. They also planned to release camera-carrying French balloons, which can range over a much larger area than the rover. The balloons can document the surface at extremely close range while compiling a good profile of the atmosphere and weather at the same time.

The double failure of the Phobos mission, combined with riots and food shortages at home, may put a brake to this ambitious Soviet plan.

The only approved U.S. Mars flight is a 1992 Mars Observer mission to map the Red Planet for minerals, water, weather, gravity, and magnetism. The probe will save money by using the structure of an RCA communications satellite and the electronics of an RCA weather satellite. Its polar orbit will allow it to remap Mars completely every 59 days, documenting any climatic changes.

The Mars Observer photos will be sharper than Landsat images of the Earth. They'll be able to resolve details as small as 1.4 meters. Perhaps they can solve the mystery of the Mars face seen in Viking orbiter photos taken in 1976 and digitally enhanced by Vincent DiPietro and Gregory Molenaar. The mile-long structure seems to be complete with two eye sockets, a nose, and a mouth. Most experts believe this complex pattern is an accident of nature, but there remains a small chance that something living built the face as a marker of some sort.

Photos of moons and planets for years have shown nothing but craters, so the Mars face is a welcome change of pace. Mars will finally provide something really interesting to investigate.

Beyond the Mars Observer mission, NASA and the White House are trying to select the next big goal. They have four options:

— Return to the Moon first, get practical experience, and mine the Moon for oxygen propellant to be used on a later expedition to Mars.

— Sprint directly to Phobos with a four-member crew, spend 20 days exploring, and return. The round-trip time would be only 440 days because the ship would use a high-energy trajectory that requires more fuel than typically flown by craft trying to make every drop count.

— Mount a full-scale Mars expedition with a crew of eight to ten people, which requires so many resources that NASA planners don't consider it a serious option at this point.

— Abandon government-run manned exploration. Taking NASA's place would be expeditions financed by sale of exclusive broadcast rights and corporate sponsorships. NASA's new role would focus on long-range research, and pure-science efforts like space astronomy.

The sprint to Phobos would resemble the Apollo program. It would produce headlines fast, but it wouldn't create an affordable long-term Mars program. The expedition to Mars is considered too expensive now to take seriously.

That leaves the Moon, used as a stepping stone to later Mars exploration. This would take longer than a sprint mission to Mars, but it also would provide permanent space assets that would slash the cost of follow-up Mars visits. If a permanent lunar base were established as part of a private expedition, we'd finally be on a course that was both affordable and self-sustaining.

With the NASA space station estimated to cost $30 billion to $80 billion, Congress is wary of endorsing new NASA goals like the Moon or Mars. The political reaction to a Mars push is very hard to predict.

At some point in the twenty-first century, however, some nation or business consortium will get Mars exploration under way. Futurists expect that an eventual Mars trade route could be maintained by two large ships that circulate between an orbiting Earthport and a Marsport near Phobos. One ship would take nine months to deliver a crew to Mars. It would stop over at Mars for 11 months and then head back to Earth. The crew it delivered would stay behind. As the first ship neared Earth, the second would be launched with another crew. The "original" crew would return on the second ship, having spent 3.2 years on Mars and a total of 1.6 years in transit between planets.

The Earthport is crucial to Earth-Mars economics. The port would be in a very high orbit—geosynchronous or even further out. That way interplanetary ships would not be fully captured by the Earth's gravity. They could leave for Mars again with much less fuel than if they'd descended to low Earth orbit or to the surface.

Trade routes need something valuable to trade. Right now, the best bet for Mars commerce would be water and fuel mined from Phobos. They could be delivered back to space stations in Earth orbit for less energy than lifting them from Earth. After the Mars Observer mission maps the planet for minerals, we may find other things worthwhile to send Earthward.

Perhaps Mars settlers will be the most important load carried. The Red Planet is smaller than Earth, but the lack of oceans makes the surface area equal to the dry area of our planet. That's a tremendous territory to explore and exploit. With self-replicating robots available to do the heavy work, a career as Mars settler could be appealing to thousands of people.

CHEAP IMPORTED STEEL—FROM ASTEROIDS

Most people think of the asteroid belt between Mars and Jupiter when asteroids are mentioned. The belt has thousands of asteroids, but they are too far away for real exploration any time in the next 20 or 30 years.

Near-Earth asteroids are much more appealing. Several dozen have been sighted, mostly by accident. Some of them come so close they require slightly less energy to reach than the Moon. The return trip to low Earth orbit is even easier because they have no gravity to escape. One potential target, an asteroid called 1982 DB, might supply metals at only *2 percent* of the energy cost of metals from the lunar surface. The drawback is a two-year round-trip compared with only six days on the lunar trade route. An asteroid mining syndicate would have its money tied up a long time before it paid off.

Scientists speculate that near-Earth asteroids should closely resemble the meteorites that hit Earth. These meteorites overwhelmingly are chondrites with a metal content from 4 to 19 percent. The metal is mostly iron combined with the nickel needed for a strong steel alloy. Meteorites also carry the platinum group of metals, including rhodium, ruthenium and palladium. They are the strategic metals needed for high-tech alloys and devices. The

United States imports almost 100 percent of its strategic metals from South Africa and the Soviet Union at a cost of $5,000 per pound.

With platinum group metals so expensive and their sources so vulnerable to disruption, importing them from space may be worthwhile. An asteroid mine might not be profitable when solely devoted to producing steel for use in orbit. But if it also produces platinum group metals as a by-product, the extra revenues could turn it into a lucrative operation.

Asteroid missions are usually proposed as automated probes rather than crewed ships. Keeping humans alive on a two-year round-trip would add enormous costs and risks, and scientists don't think a crew will be needed to supervise asteroid mining. They assume the asteroids have loose material available on the surface, so the probe wouldn't need elaborate digging equipment. The probe would scoop up the loose rocks and perform very simple extraction steps to concentrate the metals.

An easy method of concentrating the ore has been proposed by Carolyn Meinel, a space researcher who's worked on classified defense projects. She points out that many metals will react with carbon monoxide when heated to relatively low temperatures, in the range of 400 or 500 degrees. The metals and carbon monoxide form gaseous carbonyls, which then can be distilled into separate iron, nickel, and platinum containers. Heating each container then drives off the carbon monoxide to leave pure metal. The carbon monoxide that boils off can be recycled to continue the process.

The mining ship would need to bring along a start-up vat of carbon monoxide. Meinel speculates that since carbon monoxide can be extracted from the actual ore being processed, this will make up any losses of carbon monoxide during processing.

Returning the ore requires fuel, which could be created from water mined on the asteroid. The fuel needed to return each ton could be tiny, making massive iron and steel shipments quite feasible. Asteroid 1982 DB is amazingly "close" in terms of fuel needed on the return trip. The velocity required to reach low Earth orbit from 1982 DB is only 60 meters per second. By comparison, to return products from the Moon requires a speed of 3,100 meters per second. And to send materials up to low Earth orbit from the ground requires a speed of 9,000 meters per second. The 60 meters per second from 1982 DB makes returning ore shipments a snap.

Getting *to* the asteroids requires much more energy, on the order of 4,500 to 5,500 meters per second. (Going *to* the Moon also requires more energy than returning from it—about 6,000 meters per second.) The energy cost of sending the mining ship is compounded by the launch window problem. Individual asteroids are in position for the least-energy round-trip only every five years or so. Mining ships will need to shift from asteroid to asteroid to get the best launch windows for metal shipments home.

To overcome the high energy costs of sending mining ships out to meet asteroids, Meinel proposes that they not return to low Earth orbit at all. A mining ship would swing past Earth in a highly elliptical orbit. The low point would be about 100 kilometers, and the high point would be well beyond the Moon's orbit. This big loop would preserve most of the mining ship's speed. The ship would release its cargo at the low point and take on technicians to make repairs.

When the next asteroid comes into range, the mining ship would blast off at the high point of its orbit. The extra kick needed to shoot back out on another mission is much less than starting from low Earth orbit deep in the planet's gravity well.

The "Meinel maneuver" of a highly elliptical parking orbit creates the economic basis for reaching asteroids. She calculates that for every pound carried up from the Earth to support the operation, the probe will return 20 pounds of ore. This indicates that asteroid metals could be supplied to low Earth orbit for one-twentieth the cost of lifting them from Earth. The extremely good mass-payback ratio means the initial mining ships might be able to get by with bringing back loose rocks without any refining on the asteroid. The rubble could be used as radiation shielding for civilian space stations or blast and laser shielding for military bases.

Using raw dirt from asteroids as shielding simplifies the task of the initial probes. Later mining missions, after the basic technique is proved, could add the risk and complexity of on-asteroid metal extraction.

THE TWENTY-FIRST CENTURY SPACE ECONOMY

Early in the twenty-first century the space economy will be a major source of wealth. Solar power plants will be beaming extremely cheap power back to Earth, from orbit or the Moon or near Mercury. Construction crews will be in space almost constantly at work on the next solar station.

Zero-gravity research will be under way in hundreds of orbital laboratories. Precious medicines will be created and refined in space station factories. Electronic and optical devices with unrivaled powers will be available only from low-gravity production lines. Orbital repair depots for communications satellites will be staffed with scores of technicians.

Exotic metals in the platinum group will be mined from asteroids to supply the high-tech foundries on Earth. Moon bases crewed by humans and robot rovers will be churning out oxygen and aluminum. University professors will be running Moon observatories with optical telescopes, radio telescopes, and gravity wave detectors.

The service industry will be running full tilt. Spaceliners will be offering "super-saver rates" for people willing to travel during solar flares. Someone will be running a junkyard, salvaging dead satellites for spare parts. A few hotel chains will be managing small habitats for temporary workers, visiting supervisors, and affluent tourists. A recreation center is inevitable — thousands of space workers will need a place to spend their days off. They'll want a place to meet new people, eat new foods, and be entertained.

This booming space economy seems improbable to most people. It will stay "unlikely" until one critical year, when the price of a space ticket drops low enough. Perhaps the turning point will come when corporations can boost research chemists into orbit for only $50,000, or later when it will cost only $20,000. The critical point inevitably will arrive, either late this century or early the next. Then thousands of corporations and individual investors around the world will start pushing major capital into space projects. The orbital economy will undergo explosive expotential growth — each new space project drives down the launch costs for the next one. Spaceplanes launching every week are cheaper than spaceplanes on monthly missions, and spaceplanes launching every day are the cheapest of all.

Some early firms that pioneer when the risks are highest may become fabulously rich. Others may bleed to death from expenses that outstrip the possible revenues. The firms that manage to survive will have the experienced executives, trained workers, and in-place equipment to create fortunes to rival the Rockefellers, J. Paul Getty, or H. Ross Perot.

INVESTING IN THE SPACE ECONOMY

Many major corporations already devote some part of their efforts to space. Buying shares in Martin Marietta or McDonnell Douglas would be one strategy for a space investor. The drawback is that most major corporations have large nonspace activities as well, so the investment is diluted.

The pure space investments are in small firms that have a substantial risk of going bankrupt. Many of these "companies" are nothing more than one or two people with a burning desire to become space industrialists. Other new companies actually have a good assortment of key workers and at least some working capital. Here's a short list of the most promising start-up operations:

Geostar Corporation. Shrewd company executives have raised $80 million from wealthy investors, and enlisted major corporations and governments around the world as allies. Its paging and position location service likely will be a billion-dollar business by the end of the century. On the downside, its first satellite transponder died in orbit and there's no guarantee that all its future satellites will work either. Stock has been sold privately to wealthy investors and companies; a public offering is likely soon.

American Rocket Company. AMROC is the venture most likely to succeed in building a private launch system. AMROC's cost to low Earth orbit will be half that of government-surplus boosters, and will drop further as volume increases. It needs roughly $50 million to begin operations. But because investors are skeptical of completely private launch systems, it may run out of cash before it lifts off.

Spacehab Inc. The firm's shuttle expansion unit is in great demand, and NASA seems to at least tolerate the company. Spacehab needs only $15 million to $20 million from investors to develop its first module, a small amount to raise considering the customer interest already shown. The company could falter, however, if new problems ground the shuttle in the future.

Space Industries Inc. The firm's unmanned space station may be a logical place for automated factory units producing drugs and electronics. Companies could graduate from tests in the Spacehab module to full-scale production in the Space Industries' ISF. The project will cost several hundred million dollars to complete and requires a massive NASA commitment to its use, which could easily fall through.

External Tanks Corporation. The most speculative entry on this list is ETCO. The firm is little more than a set of preliminary drawings and a few wealthy backers. But the reward of turning shuttle external tanks into low-cost space habitats would be tremendous. The company is looking for corporate and individual investors willing to put up seed money.

The space economy will create major fortunes, for people quick enough to spot opportunities and cautious enough to avoid the traps. The pursuit of wealth will propel space development far faster than any bureaucrat could imagine.

A true space frontier also will be a spiritual lift. There's been no open frontier for decades, so countries feeling expansive must target their neighbors' land. Space exploration and settlement offers a modest alternative to conquest and holy wars. If space attracts only a handful of the world's passionate young dreamers, that's still a handful denied to religious fanatics, drug warlords, and revolutionary cults. Life will become a little bit safer, a tad more humane and a lot more entertaining.

Contacts for Further Information

PUBLICATIONS

Aviation Week and Space Technology
Commercial space coverage is a minor part of the magazine. Annual subscriptions are $64. McGraw-Hill Inc., 1221 Avenue of the Americas, New York, NY 10020. Phone 212-512-2000.

Final Frontier
A glossy bimonthly magazine providing excellent coverage of commercial space and general space news. One-year subscription costs $14.95. P.O. Box 20089, Minneapolis, MN 55420. Phone 612-332-3001.

Space Business News
A biweekly professional newsletter on all aspects of commercial space, with emphasis on materials processing, launch prices and government policy toward space entrepreneurs. A six-month subscription costs $227, one year for $445. Published by Pasha Publications Inc., 1401 Wilson Blvd., Arlington, VA 22209. Phone 800-424-2908 or 703-528-1244.

Space Calendar
A weekly listing of events, with profiles of active space companies. Annual subscriptions cost $79 for individuals and $139 for institutions. Space Age Publishing Company, 20431 Stevens Creek Blvd., Suite 210, Cupertino, CA 95014. Phone 408-996-9210.

Technological Progress and Space Development

A draft monograph cataloging various propulsion technologies available for lower-cost access to space, along with options for new organizational strategies. Available free from Dani Eder, Boeing Aerospace Space Station Program, Mail Stop JY-33, P.O. Box 1470, Huntsville, AL 35807.

The Prospects for Private Space Exploration

A monograph analyzing the outlook for privately funded exploration, using income from broadcast rights, corporate sponsorships and other revenues. Available for $2 and a stamped, self-addressed envelope from David Gump, 9100 White Chimney Lane, Great Falls, VA 22066.

MAJOR ORGANIZATIONS

(in general order of importance to commercial space)

Space Studies Institute

Pays for research in key technologies needed to establish space colonies and runs the biennial Conference on Space Manufacturing. Major support comes from senior associates who pledge at least $200 per year. Entry memberships are available for $15. 285 Rosedale Road, P.O. Box 82, Princeton, NJ 08540. Phone 609-921-0377.

National Space Society

A space advocacy group with active local chapters, a feisty membership and a monthly magazine, *Ad Astra.* The L5 Society merged into NSS in 1987. Dues are $30. 922 Pennsylvania Avenue SE, Washington, DC 20003. Phone 202-543-1900.

The Planetary Society

Led by astronomer Carl Sagan, the society pushes for more unmanned missions to the planets, asteroids, and comets. Introductory membership is $20. 65 North Catalina Avenue, Pasadena, CA 91106.

The Space Foundation

Founder of the Houston Space Business Roundtable; relocated in 1988 to Washington, D.C. The Space Foundation encourages the creation of space roundtables in new cities, and runs a competition for best space business plan by university students. It can be reached in care of Benner, Burnett & Coleman, 1401 NY Avenue NW, Washington, DC 20005. Phone 202-347-2414.

International Space University

Operates an eight-week summer session for graduate students in eight disciplines from engineering to law. First session was held at MIT in 1988; the second summer was scheduled for June 1989 in France. 636 Beacon Street, Suite 201, Boston, MA 02215. Phone 617-247-1987.

American Institute of Aeronautics and Astronautics

Publishes the monthly *Aerospace America,* several journals and sponsors a heavy schedule of meetings. Dues are $56. 1633 Broadway, New York, NY 10019. Phone 212-581-4300.

Young Astronaut Council

Activities for junior and senior high school students, including exchange visits with Soviet youth. 1121 Connecticut Avenue NW, Washington, DC 20036. Phone 202-682-1986.

The Challenger Center for Space Science Education

Will build five "mission sites" across the country to train teachers and elementary students in physics, chemistry, computer science and other space-related areas. 1101 King Street, #190, Alexandria, VA 22314.

Astronauts Memorial Foundation

Funded by sales of *Challenger* license plate tags in Florida, the AMF is developing a space research foundation, and a memorial at Kennedy Space Center to honor all astronauts who have died in the space program. 2121 Camden Road, Orlando, FL 32803. Phone 305-898-3737.

British Interplanetary Society
Publishes the monthly *Spaceflight* magazine and sponsors meetings on the science, history and commercialization of space. 27/29 South Lambeth Road, London SW8 1SZ. Phone 01-735-3160 for current dues in U.S. dollars.

COMPANIES ACTIVE IN COMMERCIAL SPACE

American Rocket Co.
Developing new hybrid liquid-solid rocket for low-cost launches. 847 Flynn Road, Camarillo, CA 93010. Phone 805-987-8970.

Arianespace Inc.
Offers the European Ariane rocket. 1747 Pennsylvania Avenue NW, Washington, DC 20006. Phone 202-728-9075

Battelle Columbus Laboratories
Runs a Center for Commercial Development of Space; forecasts launch demand. 505 King Avenue, Columbus, OH 43201. Phone 614-424-6424

Boeing Aerospace Co.
Space stations; heavy lift vehicles; materials processing; outreach to industry for NASA office of commercial programs. P.O. Box 3999, Seattle, WA 98124. Phone 206-773-2121

British Aerospace
Planning HOTOL single-stage-to-orbit cargo flyer. Space and Communications Division, Argyle Way, Stevenage, Hertfordshire SG1 2AS, United Kingdom; Phone 44-438-313456; telex 82130 ans BADSTE.G

Cis-Lunar Development Laboratories
Fabricating low-cost space suits using low pressure for greater flexibility.

President William C. Stone also leads ambitious underwater and cave expeditions around the world, where life support equipment is tested in rugged field conditions. 7730 Laytonia Drive, Derwood, MD 20855. Phone 301-869-1252

Defense Systems Inc.
Creator of extremely low-cost communications satellites for classified U.S. government users. DSI's low-orbit satellites use store-and-forward technology and cost about $1 million. 7903 Westpark Drive, McLean, VA 22102.

Eagle Engineering
One of the most active engineering firms aiding commercial space projects. 17629 El Camino Real, Houston, TX 77058. Phone 713-486-7292.

Earth Observation Satellite Company
Operates the Landsat remote sensing system. 4300 Forbes Boulevard, Lanham, MD 20706. Phone 301-552-0547.

External Tanks Corporation
Hopes to turn shuttle spare parts into low-cost space laboratories. 1877 Broadway, Suite 405, Boulder, CO 80302. Phone 303-444-6221.

General Dynamics
Offers the Atlas-Centaur booster. P.O. Box 85990, MZ C1-7107, San Diego, CA 92138. Phone 619-573-8000.

General Motors Research Laboratories
Plans shuttle research on new polymers, combustion physics. GM Technical Center, 30200 Mound Road, Warren, MI 48090. Phone 313-575-1265.

Geostar Corporation
Building global messaging and position-finding system. 1001 22nd Street

NW, 5th Floor, Washington, DC 20037. Phone 202-887-0870.

Global Outpost Inc.
Planning a space station devoted to tethers, refueling and on-orbit servicing of shuttles and satellites. 6836 Deer Run Drive, Alexandria, VA 22306. Phone 703-765-6235.

Great Wall Industry Corp.
Sells Long March expendable boosters. 17 Wenchang Hutong, P.O. Box 847, Beijing, China.

Instrumentation Technology Associates
Sells standard equipment for low-cost space experiments. P.O. Box 871, Exton, PA 19341. Phone 215-524-1988.

Martin Marietta
Marketing the Titan series of rockets. P.O. Box 179, Denver, CO 80201. Phone 303-977-5364.

MBB-Erno
Promoting the Sanger spaceplane and selling low-cost SPAS experiment platforms released and retrieved by the shuttle. 1655 North Fort Myer Drive, Arlington, VA 22209. Phone 703-525-9800.

McDonnell Douglas Astronautics Company
Information requests on commercial launches should be addressed to Delta II-Business Class, Dept. Y115, 5301 Bolsa Avenue, Huntington Beach, CA 92647.

McDonnell Douglas Astronautics Company
Special box number for space station information. P.O. Box 14526, St. Louis, MO 63178.

Microgravity Research Associates

Planning to create large-scale gallium arsenide in space. 1004 North Big Spring, Suite 600, Midland, TX 79701. Phone 915-684-5693.

Mitsubishi Heavy Industries

Aichi-Ken 1200 Higashi Tanaka, 485 Komaki-Shi, Japan.

Orbital Sciences Corporation

Offers upper stages to send shuttle and Titan cargo to Mars or geosynchronous orbit; also developing the Pegasus air-launched booster. 1951 Kidwell Drive, Vienna, VA 22180. Phone 703-790-0340.

Pacific American Launch Systems Inc.

Planning expendable and reusable launch vehicles. 115 Constitution Avenue, Menlo Park, CA 94025. Phone 415-323-3500.

Rockwell International

Builds shuttle orbiters, does materials processing. 12214 Lakewood Blvd., Downey, CA 90241.

Spacehab Inc.

Building an extension to the shuttle's crew compartment. 600 Maryland Avenue SW, Washington, DC. Phone 202-488-3483.

Space Industries Inc.

Building Industrial Space Facility for automated materials   processing. 711 West Bay Area Blvd., Suite 320, Webster, TX 77598. Phone 713-486-1422.

Space Services Inc.

Sells expendable launch vehicles. 7015 Gulf Freeway, Suite 140, Houston, TX 77807. Phone 713-649-1716.

SPOT Image Corp.

Sells photos from a French remote sensing satellite. 1897 Preston White Drive, Reston, VA 22091. Phone 703-620-2200.

Terra-Mar Resource Information Services Inc.

Sells systems to analyze remote sensing data. 2113 Landings Drive, Mountain View, CA 94043. Phone 415-964-6900.

3M Corporation

Developing polymer and thin film chemistry and physics. 3M Center, St. Paul, MN 55144. Phone 612-733-7229.

Wespace

A subsidiary of Westinghouse Electric that has invested in Space Industries and has conducted joint launcher studies with the American Rocket Company. Monroeville, PA.

Wyle Laboratories

Proposes a commercial lab-for-hire on the space station. 7800 Governors Drive, Huntsville, AL 35807. Phone 709-452-9133.

MAJOR NASA OFFICES

NASA

Office of Commercial Programs, Code I, Washington, DC 20546. Phone 202-453-1123.

NASA

Office of Exploration, Code Z, Washington, DC 20546. Phone 202-453-8928.

NASA
Office of Space Station, Code S, Washington, DC 20546. Phone 202-453-2015.

NASA
Space Station Program Office, Parkridge III, 10701 Parkridge Blvd., Reston, VA 22091. Phone 703-487-7000.

NASA
Public Affairs, Code LFD-2, Washington, DC 20546. Phone 202-453-8400.

NASA - Johnson Space Center
Public Information, Code AP3, Houston, TX 77058. Phone 713-483-5111.

NASA - Kennedy Space Center
Public Information, Code PA-PIB, Kennedy Space Center, FL 32899. Phone 305-867-2468.

NASA - Marshall Space Flight Center
Public Information, Code CA10, Marshall Space Flight Center, AL 35812. Phone 205-453-0034.

NASA-SUPPORTED COMMERCIAL SPACE CENTERS

Center for Advanced Materials
Battelle Columbus Laboratories, 505 King Avenue, Columbus OH 43201. Phone 614-424-7240. Dr. Francois R. Mollard, director.

Center for Advanced Space Propulsion
University of Tennessee Space Institute, P.O. Box 1385, Tullahoma, TN

37388. Phone 615-454-9294. Dr. Fred Speer, director.

Center for Autonomous & Man-Controlled Robotic & Sensing Systems
ERIM, P.O. Box 8618, Ann Arbor, MI 48107.   Phone 313-994-1200
x2457. Dr. Charles Jacobus, director.

Center for Bioserve Research
University of Colorado-Boulder, School of Aerospace Engineering
Sciences, Campus Box 429, Boulder, CO 80309. Phone 303-492-7613. Dr.
Marvin Luttges, director.

Center for Cell Research
Pennsylvania State University, 465 North Frear Laboratory, University
Park, PA 16802. Phone 814-865-9182. Dr. Wesley Hymer, director.

Center for Development of Commercial Crystal Growth in Space
Center for Advanced Materials Processing, Clarkson University, Potsdam
NY 13676. Phone 315-268-2336. Dr. William Wilcox, director.

Center for Macromolecular Crystallography
University of Alabama-Birmingham, THT-Box 79, University Station,
Birmingham, AL 35294. Phone 205-934-5329. Dr. Charles Bugg, director.

Center for Mapping
Ohio State University, 1958 Neil Avenue, Columbus, OH 43210. Phone
614-292-6642. Dr. John Bossler, director.

Center for Materials for Space Structures
Case Western Reserve University, School of Engineering, Department of
Materials Science and Engineering, 10900 Euclid Avenue, Cleveland, OH
44106. Phone 216-368-4222. Dr. John Wallace, director.

Center for Space Automation and Robotics
University of Wisconsin-Madison, 1357 University Avenue, Madison WI 53706. Phone 608-262-5542. Dr. John Bollinger, director.

Center for Space Power
Space Research Center, Zachary Bldg. Room 218, Texas A&M University, College Station, TX 77843. Phone 409-845-8768. Dr. Alton Patton, director.

Center for Space Processing of Engineering Materials
Vanderbilt University, Box 6309-Station B, Nashville, TN 37235. Phone 615-322-7047. Dr. Robert Bayuzick, director.

Center for Space Vacuum Epitaxy
University of Houston, Science & Research Bldg. I, Houston, TX 77204. Phone 713-749-3701. Dr. Alex Ignatiev, director.

Consortium for Materials Development in Space
University of Alabama-Huntsville, Research Institute Bldg., Hunstville, AL 35899. Phone 205-895-6620. Dr. Charles Lundquist, director.

ITD Space Remote Sensing Center
Bldg. 1103, Suite 118, John Stennis Space Center, MS 39529. Phone 601-688-2509. Dr. George May, director.

Space Power Institute
Auburn University, 231 Leach Center, Auburn, AL 36849. Phone 205-826-5894. Dr. Raymond Askew, director.

Index

Advanced Launch System, 50, 52-53
Alabama, University of, 144
Alcoa, 144
Aldrich, Arnold, 14
Allen, Joseph, 167
Allied Signal, 144
American Mobile Satellite Consortium, 106
American Rocket Company, 3, 31-35
American Telephone & Telegraph, 91, 145
Amoco Chemicals, 144
anti-matter engines, 86
Apollo, Project, 1
Arabsat, 122
Arianespace, 41-42, 122
Armco, 144
Asahi, 147
asteroids, 199-201
Atlas-Centaur rocket, 40
Autodesk Inc., 181

Battelle Columbus Laboratories, 144
Bekey, Ivan, 83
Bell, Alexander Graham, 184
Bennett, James, 31
Beyer, Larry, 190
BioCryst Ltd., 139
BMW, 142

Boeing, 23, 52, 144, 174
Boisjoly, Roger, 9-10, 20
Booz, Allen & Hamilton, 143
Bristol Aerospace, 37
British Aerospace, 67, 69, 71
Bugg, Charles, 139

Canada, 170
cancer research, 5
centers for commercial development of space, 143
Challenger, 1, 9
Chow, Chris, 132
Chu, Paul, 110
Cis-Lunar Development Laboratories, 194
C. Itoh, 146
Clarkson University, 145
Claybaugh, William, 137
CNES, 101, 114
Comsat, 100
Connestoga I, 37
Cook, Paul, 129
Coopers and Lybrand, 143
Coopersmith, Joshua, 184
Crew Emergency Return Vehicle, 176
Criswell, David, 189

Daimler-Benz, 142
Deere & Co., 136, 144
Defense Advanced Research
  Projects Agency, 35-37, 88
DeForest, Lee, 183-184
Delta rocket, 40
Diamandis, Peter, 6
direct broadcast satellites, 93
Dow Chemical, 133
Dr. Zhivago, 13
Dunstan, James, 194

Earth Observation Satellite Co., 154
Eldred, Charles, 62
Electro-Optek Corp., 145
electrophoresis, 112, 137
Energia booster, 43
Engelhard Corp., 144
ENIAC computer, 183
erythropoeitin, 112
Escher, William, 66-67
Eudy, Robert, 19
European Space Agency, 73, 170
expendable launch vehicles, 27
external tank, 51, 177-182, 193
External Tanks Corp., 179

Faget, Maxime, 166
Fletcher, James, 19
Freedom (NASA's proposed space
  station) 48-49, 168-177
free electron lasers, 78-79
Fuji Heavy Industries, 74
fusion, 120

Gatos, Henry, 116
General Dynamics, 40, 52
General Electric, 174
General Motors, 127, 144
GeoDecisions, 157

GeoInformation Systems, 157
Geostar Corp., 3, 97-108
geosynchronous orbit, 91
Glenn, John, 1
Global Outpost, 181
Global Positioning System, 40
Globesat Inc., 35
GTE Satellite Corp., 102, 144
Guilford Transportation Industries,
  100

Halley's comet, 16
Hannah, David Jr., 38
Hardy, George, 10
Hawley, Todd, 6
Hercules Corp., 36
Hermes shuttle, 73
high-definition television, 94
Hitachi, 146
Hoffman, Hans, 142
HOPE spaceplane, 74
HOTOL, 67-71
House Science and Technology
  Committee, 21
Houston Industries, 38
Houston, University of, 145
Huang Zuo Yi, 44
Hudson, Gary, 38
Hughes, Brian, 115
Hutson Fertilizer Co., 154

IBM, 144
II-IV Inc., 144
Industrial Space Facility, 166
Inmarsat, 42
Institute for Technology Develop-
  ment, 144, 154
International Paper Co., 144
International Space University, 6, 47
INTOSPACE, 44, 110, 142

Johnson Controls, 145
Johnson & Johnson, 113
joint endeavor agreement, 116, 126

Kawasaki Heavy Industries, 74, 146
Keyworth, George, 57, 101
King, Charles, 193
Kobe Steel, 146
Koopman, George, 32-33
Kyocera, 146

laser beam propulsion, 77-82
Lockheed Missiles and Space Co.,
   174
Long March booster, 43
Lund, Robert, 10, 20

magnetic launchers, 85
Martin Marietta, 39-40, 52, 88
Mason, Jerald, 10
mass drivers, 85
MBB-Erno, 72
McDonnell Douglas, 40-41, 52, 112-
   114, 137, 166, 174
Meinel, Carolyn, 200
Mendel, Wendell, 192
Microgravity GmbH, 143
Microgravity Research Associates,
   115-120
Miller, John, 19
Mir (Soviet space station) 23, 114,
   140, 148
Mitsubishi Heavy Industries, 74, 146
Mitsui, 146
Morton Thiokol, 9-10, 19-20, 28, 37
Mulloy, Larry, 10
Munitions Control Act, 43
Myrabo, Leik, 81

National Aerospace Plane, 50, 53-62

NAVSTAR, 3, 105, 107
Nippon Electric, 146
Nissan Motors, 74, 146
nose cone, 11-12

Ohio State University, 145
O'Neill, Gerard K., 97, 99-102, 104-
   105
Orbital Sciences Corp., 35-37
orbital transfer vehicle, 163-164
Orient Express, 53
O-rings, 9-11, 14, 19-20

Pacific American Launch Systems,
   38-39
Palestine Liberation Organization,
   122
Pegasus, 36
Penzo, Paul, 83-84
Perkin-Elmer Co., 145
Phobos, 195
Podsiadly, Chris, 125
Poletayev, Dmitry, 43
PPG Industries, 144
Pratt & Whitney, 59
Proton booster, 42-43

Qaddafi, Muamar, 122
Qualcom, 106

Ramsland, Russell Jr., 115-120
Richards, Bob, 6
Rockwell International, 13, 59, 145,
   174
Rogers Commission, 21
Rogers, T.F., 168, 180
Rolls Royce, 69
Rothblatt, Martin, 100, 104-105

Sahm, Peter, 141

Salyut 1, 22
Sänger spaceplane, 71-73
Scobee, Dick, 14
Scott, Michael, 28, 31
scramjet, 60
Skylab, 6, 22, 116, 173
shuttle, 2, 16
shuttle C, 51
shuttle external tank. *See* external
  tanks
Simon, William, 105
Slayton, Deke, 38
Sloyer, Jack Jr., 137
Smith, Mike, 14-15
Snap-On Tools, 145
Society Expeditions, 38
solar power satellites, 185-188
Sony, 97
South Atlantic Anomaly, 175
Spar Aerospace, 171
Spacehab Inc., 165
Space Industries, 166
space services development agree-
  ment, 101, 167
Space Services Inc., 37-38
space stations. See *Freedom; Mir;*
  External Tanks Corp.; Global Out-
  post; *Skylab*
Space Studies Institute, 85, 105, 107,
  179, 185
space tourists, 47

Starstruck Inc., 28-31
Steurer, Wolfgang, 188
Stone, William, 194
Strategic Defense Initiative, 78-80,
  187
Sumitomo, 146

Talay, Theodore, 62
tethers, 82-85
Thompson, Arnold, 9
3M, 114, 125-134, 166
tiles, 17-18
Titan rocket, 39
Toshiba, 146

University Corporation for Atmospheric Research, 180
Vanderbilt University, 144
vanHoften, Ox, 131
Volkswagen, 142

Waldron, Robert, 189
Westinghouse Electric Co., 35, 86,
  167
Williams, Robert, 57
Wisconsin, University of, 144
Woodcock, Gordon, 191
Wright, Orville and Wilbur, 23
Wyle Laboratories, 144

X-30. *See* National Aerospace Plane

About the Author

DAVID P. GUMP has experience in space commerce from the vantage point of marketing and publishing. His credits include the creation in 1983 of *Space Business News,* the major newsletter serving the space commerce field. He later supervised *Military Space,* another Pasha Publications newsletter for Pentagon space programs.

Mr. Gump organized two major space conferences for Pasha Publications that drew hundreds of executives from firms around the world. The seminars outlined the business opportunities in the shuttle program and the new space station.

He has also served as director of business development for Geostar Messaging Corporation, which is planning a satellite-based communications system for voice, data, and facsimile.

Mr. Gump, who lives in Great Falls, Virginia, holds a B.A. from Macalester College and an M.S. from Northwestern University. He is currently a consultant, advising firms in space commerce and other high-tech fields.